# Beneficial Microbes for Sustainable Crop Production

# Beneficial Microbes for Sustainable Crop Production

Guest Editors

**Vlad Stoian**
**Roxana Vidican**

MDPI

Basel • Beijing • Wuhan • Barcelona • Belgrade • Novi Sad • Cluj • Manchester

*Guest Editors*

Vlad Stoian
Department of Microbiology
University of Agricultural
Sciences and Veterinary
Medicine Cluj-Napoca
Cluj-Napoca
Romania

Roxana Vidican
Department of Microbiology
University of Agricultural
Sciences and Veterinary
Medicine Cluj-Napoca
Cluj-Napoca
Romania

*Editorial Office*
MDPI AG
Grosspeteranlage 5
4052 Basel, Switzerland

This is a reprint of the Special Issue, published open access by the journal *Agriculture* (ISSN 2077-0472), freely accessible at: https://www.mdpi.com/journal/agriculture/special_issues/W1F7L85ZA0.

For citation purposes, cite each article independently as indicated on the article page online and as indicated below:

| Lastname, A.A.; Lastname, B.B. Article Title. *Journal Name* **Year**, *Volume Number*, Page Range. |
|---|

**ISBN 978-3-7258-6360-0 (Hbk)**
**ISBN 978-3-7258-6361-7 (PDF)**
**https://doi.org/10.3390/books978-3-7258-6361-7**

Cover image courtesy of Vlad Stoian

# Contents

# About the Editors

**Vlad Stoian**

Vlad Stoian is an Associate Professor at the Department of Microbiology, Faculty of Agriculture, University of Agricultural Sciences and Veterinary Medicine (USAMV) Cluj-Napoca, Romania. His research focuses on microbial communities in soils, their reaction to different technological inputs, and patterns of activity dynamics. Another contribution was to improve the extraction of parameters for mycorrhizal colonization through the application of a pattern comparison methodology.

**Roxana Vidican**

Roxana Vidican is a Professor at the Department of Microbiology, Faculty of Agriculture, University of Agricultural Sciences and Veterinary Medicine (USAMV) Cluj-Napoca, Romania. Her work combines different subjects of general and applied microbiology in agroecosystems. The research carried out in the field was focused on microorganisms' reaction to stress factors, salinity, heavy metals, fertilizer and amendment applications, respectively, and the native and induced mycorrhizal colonization potential in different species.

*agriculture*

MDPI

*Editorial*

# Beneficial Microbes for Sustainable Crop Production

**Vlad Stoian and Roxana Vidican ***

Department of Microbiology, Faculty of Agriculture, University of Agricultural Sciences and Veterinary Medicine of Cluj-Napoca, 3-5 Calea Mănăştur, 400372 Cluj-Napoca, Romania; vlad.stoian@usamvcluj.ro
* Correspondence: roxana.vidican@usamvcluj.ro

Microbial communities represent a major component of cultivated soils and are responsible for the successful production of crops [1–3]. The evolution of soil microbiomes in the current agronomic context depends on both soil type and fertility, as well as the cultivated species and technologies, and has demonstrated large variability under the impact of climate change [4–7]. Numerous and diverse microorganisms are involved in securing nutrients for their host and in sustaining their optimum growth and development processes [8–10]. Soil inoculation with microbial consortia represents a viable solution to increase the amount of nutrients required by crops, and their protection against biotic and abiotic stresses, alongside ensuring stability in soil fluxes and organic matter recycling [11–14].

The Special Issue "Beneficial Microbes for Sustainable Crop Production" is a collection of eight manuscripts that explore the complex interactions between soil, plants, and microbial communities. We want to thank all the contributors for their studies, which were performed in field and controlled environments. All these studies deepen the understanding on microbial performance and functions and sustain the development of future agronomic strategies and applications.

A pot study with six arbuscular mycorrhizal species was conducted on maize [15], with the aim of analyzing the effect of inoculation on multiple plant and soil parameters. The authors assessed plant growth, leaf gas exchange and sugar levels, mycorrhizal colonization rate in roots, glomalin-related soil proteins, and conducted an analysis of *SWEET* gene expression. Their findings indicate significant species-specific variation in terms of root colonization, improvements in plant traits, and glomalin-related soil protein fractions. The same species-specific effect was observed in reprogrammed *ZmSWEET* expression, with core *ZmSWEET* genes mediating sugar accumulation to support symbiosis. The study provides a foundation for further exploration of the interaction between arbuscular mycorrhizal fungi and *SWEET* genes.

Two commercial cultivars (Pedrosillano and Blanco Lechoso) and twenty *Cicer arietinum* L. (chickpea) germplasm were tested against seven *Mesorhizobium* strains for cultivar-specific symbiotic performance [16]. The research was conducted by the authors under growth-chamber and greenhouse conditions, in the presence and absence of drought stress. During the initial screening, three elite strains (ISC11, ISC15, and ISC25) showed superior symbiotic performance and nitrogen activity. The effect of drought was significantly mitigated in several strain–cultivar combinations, maintaining over 70% of shoot biomass compared to controls. Based on the whole-genome sequencing results, diverse taxonomic affiliations were observed—*Mesorhizobium ciceri* (ISC11), *Mesorhizobium mediterraneum* (ISC15), and a potential novel species (ISC25). The findings of the authors show the potential of targeted rhizobial inoculants optimized for chickpea cultivars with the aim of improving crop performance under water-limiting conditions.

The bacterial populations in five turmeric (*Curcuma*) species were analyzed in field and pot trials [17]. The bacterial populations recorded in the rhizosphere, stems, and leaves showed variations across growth stages, with the highest values reached in the rhizosphere in both types of experiments. A compositional distinction between leaf-associated bacterial populations and those from rhizosphere and stems was observed. The results showed that *Curcuma* growth stage and species have a significant effect on the bacterial community structure. The findings of the authors can facilitate a better understanding of plant–bacteria dynamics and the optimization of microbial inoculation strategies for the development of sustainable agricultural practices.

Seedling P uptake and mycorrhizal colonization in rice and pearl millet was assessed under different arbuscular mycorrhizal fungi inoculation and water regime conditions [18]. The results presented by the authors showed that two new inoculants ($I_2$ and $I_3$) had higher propagule numbers and an increase in the infection rate compared to the control seedlings, while another inoculant ($I_1$) improved root transversal area and shoot growth parameters. All inoculants showed lower infection rates than the indigenous arbuscular mycorrhizal fungi from upland Andosol. For pearl millet, a higher P uptake and shoot dry weight was observed in conditions involving a higher infection rate. The study showed the importance of inoculants for seedling establishment and highlighted the increased P uptake through mycorrhizal mediation in pearl millet than in rice.

An extensive biochemical characterization of *Phytobacter* sp. isolate, WL65, yielding genomic insights, was conducted to accurately identify the species and their functional potential for plant growth promotion [19]. In addition, the research predicted secondary metabolic pathways through genomic analyses and the potential use of the isolate as biofertilizer for rice cultivation. Based on whole-genome analysis, the identity of *Phytobacter palmae* WL65 was confirmed, as well as the presence of genes for nitrogen fixation, phosphate solubilization, IAA biosynthesis, siderophore production, and biofilm formation. The application of the isolate in rice increased plant growth and yield, also improving soil fertility. The findings of the study sustain the potential of WL65 as biofertilizer and improve the understanding of plant growth-promotion rhizobacterium mechanism.

Plant growth-promoting rhizobacteria and hydrogel (potassium polyacrylate) were analyzed on *Brassica napus* L. under drought conditions [20]. *Serratia plymuthica* was selected from six candidates, based on in vitro and pot experiments, where it showed an enhancement in plant biomass, shoot length, and the number of internodes. The application of hydrogel showed no adverse effects on seed viability and tested bacterial strains. Field experiments showed an increase in the number of siliques, yield, plant height, and the number of branches after the application of *S. plymuthica*. The application of hydrogel (solely or with bacterium) to the soil delayed seedling emergence and diminished the growth-promoting effect of *S. plymuthica*. The findings of the study indicate the potential use of *S. plymuthica* as an ingredient for seed coatings.

Research on the reduction in glucosinolates levels in transgenic rapeseed and the presence of arbuscular mycorrhiza in the rhizosphere was performed to test if the formation of specific fungal structures would be possible [21]. After the establishment of the symbiosis, the main objective was to identify the beneficial effects of the association on seed yield. The conclusion of the study was that inoculation with arbuscular mycorrhiza of transgenic rapeseed plants expressing the *Thkel1* gene improves the seed yield and fatty acid composition of oilseed. However, the reduction in glucosinolates levels is not enough to sustain the formation of specific arbuscular mycorrhiza structures—arbuscules and vesicles. The results of the research represent new opportunities in agrobiotechnology for the application of arbuscular mycorrhizal fungi as biofertilizers for Brassicaceae crops.

The effects of individual and combined inoculation with microorganisms on the polyphenol content in Okra (*Abelmoschus esculentus* L.) were analyzed for the detection of synergistic combinations [22]. The tested microorganism had generally positive effects on plant growth and yield, also leading to an increase in the glutathione-S-transferase enzyme activity of leaves. Individual inoculation showed higher enzyme activities compared to combined treatment. The treatment, involving a combination of *F. mosseae* and *Streptomyces* strain K61, led to a significant increase in the coumaric acid content, while *Aureobasidium* strain DSM 14950 showed a similar effect on quercetin and quercetin-3-diglucoside. The findings of the study showed that targeted inoculation can have a selective influence on the content of specific polyphenol compounds.

**Author Contributions:** Conceptualization, V.S. and R.V.; writing—original draft preparation, V.S. and R.V.; writing—review and editing, V.S. and R.V. All authors have read and agreed to the published version of the manuscript.

**Data Availability Statement:** No new data were created or analyzed in this study. Data sharing is not applicable to this article.

**Conflicts of Interest:** The authors declare no conflicts of interest.

## References


1. Dincă, L.C.; Grenni, P.; Onet, C.; Onet, A. Fertilization and Soil Microbial Community: A Review. *Appl. Sci.* **2022**, *12*, 1198. [CrossRef]
2. Chen, Q.; Song, Y.; An, Y.; Lu, Y.; Zhong, G. Soil Microorganisms: Their Role in Enhancing Crop Nutrition and Health. *Diversity* **2024**, *16*, 734. [CrossRef]
3. Philippot, L.; Chenu, C.; Kappler, A.; Rillig, M.C.; Fierer, N. The Interplay between Microbial Communities and Soil Properties. *Nat. Rev. Microbiol.* **2024**, *22*, 226–239. [CrossRef] [PubMed]
4. Shah, A.M.; Khan, I.M.; Shah, T.I.; Bangroo, S.A.; Kirmani, N.A.; Nazir, S.; Malik, A.R.; Aezum, A.M.; Mir, Y.H.; Hilal, A.; et al. Soil Microbiome: A Treasure Trove for Soil Health Sustainability under Changing Climate. *Land* **2022**, *11*, 1887. [CrossRef]
5. Jansson, J.K.; McClure, R.; Egbert, R.G. Soil Microbiome Engineering for Sustainability in a Changing Environment. *Nat. Biotechnol.* **2023**, *41*, 1716–1728. [CrossRef] [PubMed]
6. Chernov, T.I.; Semenov, M.V. Management of Soil Microbial Communities: Opportunities and Prospects (a Review). *Eurasian Soil Sci.* **2021**, *54*, 1888–1902. [CrossRef]
7. Stoian, V.; Vidican, R.; Florin, P.; Corcoz, L.; Pop-Moldovan, V.; Vaida, I.; Vâtcă, S.-D.; Stoian, V.A.; Pleșa, A. Exploration of Soil Functional Microbiomes—A Concept Proposal for Long-Term Fertilized Grasslands. *Plants* **2022**, *11*, 1253. [CrossRef] [PubMed]
8. Nagrale, D.T.; Chaurasia, A.; Kumar, S.; Gawande, S.P.; Hiremani, N.S.; Shankar, R.; Gokte-Narkhedkar, N.; Renu; Prasad, Y.G. PGPR: The Treasure of Multifarious Beneficial Microorganisms for Nutrient Mobilization, Pest Biocontrol and Plant Growth Promotion in Field Crops. *World J. Microbiol. Biotechnol.* **2023**, *39*, 100. [CrossRef] [PubMed]
9. Mohamed, H.I.; Sofy, M.R.; Almoneafy, A.A.; Abdelhamid, M.T.; Basit, A.; Sofy, A.R.; Lone, R.; Abou-El-Enain, M.M. Role of Microorganisms in Managing Soil Fertility and Plant Nutrition in Sustainable Agriculture. In *Plant Growth-Promoting Microbes for Sustainable Biotic and Abiotic Stress Management*; Mohamed, H.I., El-Beltagi, H.E.-D.S., Abd-Elsalam, K.A., Eds.; Springer International Publishing: Cham, Switzerland, 2021; pp. 93–114. ISBN 978-3-030-66587-6.
10. Singh, S.K.; Wu, X.; Shao, C.; Zhang, H. Microbial Enhancement of Plant Nutrient Acquisition. *Stress Biol.* **2022**, *2*, 3. [CrossRef] [PubMed]
11. Kaushal, M.; Devi, S.; Kumawat, K.C.; Kumar, A. Microbial Consortium: A Boon for a Sustainable Agriculture. In *Climate Change and Microbiome Dynamics: Carbon Cycle Feedbacks*; Parray, J.A., Ed.; Springer International Publishing: Cham, Switzerland, 2023; pp. 15–31. ISBN 978-3-031-21079-2.
12. Wu, D.; Wang, W.; Yao, Y.; Li, H.; Wang, Q.; Niu, B. Microbial Interactions within Beneficial Consortia Promote Soil Health. *Sci. Total Environ.* **2023**, *900*, 165801. [CrossRef] [PubMed]
13. Aguilar-Paredes, A.; Valdés, G.; Nuti, M. Ecosystem Functions of Microbial Consortia in Sustainable Agriculture. *Agronomy* **2020**, *10*, 1902. [CrossRef]
14. Vishwakarma, K.; Kumar, N.; Shandilya, C.; Mohapatra, S.; Bhayana, S.; Varma, A. Revisiting Plant–Microbe Interactions and Microbial Consortia Application for Enhancing Sustainable Agriculture: A Review. *Front. Microbiol.* **2020**, *11*, 560406. [CrossRef] [PubMed]

15. He, G.-X.; Zheng, F.-L.; Zou, Y.-N.; Gao, X.-B.; Wu, Q.-S.; Guo, C. Mycorrhizal Regulation of Core ZmSWEET Genes Governs Sugar Accumulation in Maize. *Agriculture* **2025**, *15*, 1790. [CrossRef]
16. Camacho, M.; Vaccaro, F.; Brun, P.; Ollero, F.J.; Pérez-Montaño, F.; Negussu, M.; Martinelli, F.; Mengoni, A.; Rodriguez-Navarro, D.N.; Fagorzi, C. Selection and Characterisation of Elite *Mesorhizobium* spp. Strains That Mitigate the Impact of Drought Stress on Chickpea. *Agriculture* **2025**, *15*, 1694. [CrossRef]
17. Raharja, N.I.; Hossain, M.A.; Akamine, H. Microbial Population in Curcuma Species at Different Growth Stages. *Agriculture* **2025**, *15*, 1092. [CrossRef]
18. Y, P.; Kamoshita, A. Arbuscular Mycorrhizal Fungi Inoculation and Water Regime Effects on Seedling P Uptake by Rice and Pearl Millet. *Agriculture* **2025**, *15*, 753. [CrossRef]
19. Thamvithayakorn, P.; Phosri, C.; Vishal, V.; Suwannasai, N. Characterization and Whole-Genome Sequencing of *Phytobacter palmae* WL65, a Plant Growth-Promoting Rhizobacterium First Isolated from Rice Rhizosphere Soil in Thailand. *Agriculture* **2025**, *15*, 707. [CrossRef]
20. Dąbrowska, G.B.; Krauklis, D.; Kulasek, M.; Nocny, M.; Antoszewski, M.; Mierek-Adamska, A.; Kaliska, B. Synthetic Hydrogel Dilutes *Serratia plymuthica* Growth—Promoting Effect on *Brassica napus* L. Under Drought Conditions. *Agriculture* **2025**, *15*, 142. [CrossRef]
21. Nicolás, C.; Calvo-Polanco, M.; Poveda, J.; Alonso-Ramírez, A.; Ascaso, J.; Arbona, V.; Hermosa, R. The Presence of Arbuscular Mycorrhizal Fungi in the Rhizosphere of Transgenic Rapeseed Overexpressing a *Trichoderma* Thkel1 Gene Improves Plant Development and Yield. *Agriculture* **2024**, *14*, 851. [CrossRef]
22. Almuslimawi, A.A.A.; László, L.; Sahad, A.L.; Yousif, A.I.A.; Turóczi, G.; Posta, K. Preliminary Results of the Impact of Beneficial Soil Microorganisms on Okra Plants and Their Polyphenol Components. *Agriculture* **2024**, *14*, 776. [CrossRef]

agriculture

*Article*

# Mycorrhizal Regulation of Core *ZmSWEET* Genes Governs Sugar Accumulation in Maize

Guang-Xia He [1,2,3,4], Feng-Ling Zheng [1], Ying-Ning Zou [1], Xiu-Bing Gao [1,2,3,4], Qiang-Sheng Wu [1,*] and Can Guo [1,2,3,4,*]

1 Hubei Key Laboratory of Spices & Horticultural Plant Germplasm Innovation & Utilization, College of Horticulture and Gardening, Yangtze University, Jingzhou 434025, China; 2024710911@yangtzeu.edu.cn (G.-X.H.); gxb527@163.com (X.-B.G.)
2 Tea Research Institute, Guizhou Province Academy of Agricultural Science, Guiyang 550006, China
3 Guizhou Key Laboratory of Agricultural Microbiology, Guizhou Province Academy of Agricultural Science, Guiyang 550006, China
4 Zunyi Comprehensive Field Scientific Observation and Research Station of the Ministry of Agriculture and Rural Affairs, Zunyi 564100, China
* Correspondence: wuqiangsheng@yangtzeu.edu.cn (Q.-S.W.); cancan.good@163.com (C.G.)

**Abstract:** Mycorrhizal symbiosis relies on the host's supply of carbohydrates, while sugar transport within plants is governed by the *SWEET* sugar transporter family. Although the symbiotic association between arbuscular mycorrhizal fungi (AMF) and maize is critical for its growth and sugar regulation, different AMF species have varying impacts on the host. The aim of this study was to analyze the effects of inoculating six different AMF species [*Diversispora epigaea* (*De*), *Rhizophagus intraradices* (*Ri*), *Paraglomus occultum* (*Po*), *Entrophospora etunicata* (*Ee*), *Glomus heterosporum* (*Gh*), and *Funneliformis mosseae* (*Fm*)] on plant growth, leaf photosynthetic capacity, glomalin-related soil protein content, leaf sugar content, and *SWEET* gene expression of maize under potted conditions for two months. AMF species colonize maize roots and showed significant species-specific variation, where *Ri* and *Fm* colonized treatment had the greatest rates (66~68%). All six fungi significantly increased biomass and stem diameter, with *Ee* treatment yielding the thickest stems, and enhanced leaf photosynthetic performance and glomalin-related soil protein fractions to some extent, with species-specific enhancements. All AMF species in particular significantly increased leaf sucrose; all except *Ri* treatment significantly increased fructose; while only *Po* and *Fm* treatments significantly increased glucose. AMF inoculations consistently upregulated the expression of *ZmSWEET1b/3a/3b/4a/4b/14a* and *16* genes, consistently downregulated the expression of *ZmSWEET6b/11b/12a/13a/13b/13c* and *17b* genes, and induced treatment-specific regulation in the other gene expression. Root AMF colonization clustered with sugars and specific *ZmSWEETs*, with *ZmSWEET4a/15b* and *14b* central to sucrose/glucose based on principal component analysis, indicating that these genes have specific regulatory effects in response to AMF treatments. In short, AMF inoculation reprogrammed *ZmSWEET* expression in a species-specific manner, with core *ZmSWEET* genes mediating sugar accumulation to support symbiosis.

**Keywords:** glucose; sucrose cleavage; sugar transporter genes; symbiosis; *Zea mays*

## 1. Introduction

Maize (*Zea mays* L.) is one of the world's most important and widely cultivated crops [1], leading global agricultural production in terms of annual yield, cultivated area, and economic value. Among all cereal grains, the importance of maize ranks second only to wheat and rice [2]. In China, the sown area of maize reached 44.7407 million hectares (ha) in 2024, with a yield of 6.5917 t/ha and a total output of 294.917 million tons, ranking first globally in maize production [3]. Arbuscular mycorrhizal fungi (AMF) could inhabit the maize rhizosphere, then colonizing the roots to establish mycorrhizal symbiosis [4]. Following symbiosis establishment, AMF hyphae could penetrate root cortical cells to form specialized structures including arbuscules (nutrient-exchange hub) and vesicles (storage and propagation) [5]. Within this symbiotic relationship, the host plant transfers photosynthates to the AMF to support fungal growth [6], while the AMF extends the absorption range of the host roots through extraradical hyphae to facilitates the transport of water and mineral nutrients (such as phosphorus and copper) from the soil to the host plant [7,8].

AMF have many important physiological functions in maize. AMF colonization causes morphological changes in maize roots, thus enhancing nutrient uptake [9]. AMF also significantly improve phosphorus (P) absorption, which is vital for maize growth, especially in P-limited soils [10]. AMF secrete glomalin, an insoluble glycoprotein that binds to soil humic-mineral complexes, forming glomalin-related soil proteins (GRSP) [11]. The GRSP can stabilize soil aggregates and contribute to soil organic carbon sequestration [12–14]. Inoculation with specific AMF, such as *Rhizophagus irregularis*, has been shown to increase the biomass and nutrient content of maize, enhancing the uptake of essential minerals like calcium, magnesium, and sulfur [15]. Mycorrhizal associations lead to denser and more branched root systems in maize, which facilitate better nutrient and water acquisition and plant performance under resource-limited conditions [15,16]. Although AMF are generally beneficial for maize, the specific type of AMF can lead to varying degrees of effectiveness in nutrient uptake and root growth [17–19]. Different indigenous strains of AMF significantly influenced maize yield and productivity, with co-inoculation of *Funneliformis geosporum*, *Glomus caledonius*, and *Rhizophagus intraradices* combined with reduced mineral fertilizers yielding a 62.5% increase in grain yield [17]. In ferruginous soil in Northern Benin, different AMF strains, particularly *Funneliformis mosseae*, significantly enhanced maize growth and yield [20]. This variability suggests that choosing appropriate AMF strains could optimize maize cultivation under different environmental conditions.

Sugar, as a substrate for the production of primary and secondary metabolites, is a crucial fundamental carbon source supporting plant morphogenesis and growth [21]. It participates in energy metabolism and signal transduction to sustain plant development [22]. The distribution of photoassimilates in plants relies on sugar transporters to control assimilation and the efficient transport of sugars to various tissues and cells [23]. The SWEET (Sugars Will Eventually Be Exported Transporters) family in plants has been identified as an unusually conserved form of sugar transporters, present in nearly all plants [24]. SWEET proteins can transport glucose, sucrose, fructose, and galactose, and are involved in sugar transport and accumulation within plants [25]. The observed increase in sucrose and glucose concentrations in AMF-colonized citrus seedlings reflects a reciprocal relationship between sugar accumulation and fungal colonization: AMF colonization triggers upregulation of plant photosynthetic genes and enhances carbon fixation, increasing root sugar availability; and these elevated sugar levels then sustain fungal growth and promote further colonization [26]. The interaction between AMF and host plants involves complex molecular signaling and transcriptional changes, including the regulation of *SWEET* genes, which are crucial for nutrient exchange and symbiotic maintenance [27]. The colonization of AMF could lead to significant transcriptional reprogramming of the *SWEET* sugar transporter family in plants like potatoes, with changes observed in 22 out of 35 *SWEET* genes in roots [28]. During AM symbiosis in soybeans, the *SWEET* transporter *GmSWEET6* was upregulated, enhancing sucrose transfer towards AMF, which is crucial for maintaining the symbiotic association [29]. In *Medicago truncatula*, the *SWEET1b* transporter was upregulated in arbuscule-containing cells, facilitating glucose transport across the peri-arbuscular membrane, which is vital for arbuscule maintenance [30].

Current studies have shown that AMF are beneficial for promoting sugar synthesis (especially sucrose synthesis) in plants [31], but the effects of different AMF species on the host vary. Does this selectivity differ between species of AMF? Do certain AMFs have a greater tendency to regulate specific *SWEET* genes? So, the purpose of this study is to analyze the effects of six different AMF species on maize growth, glomalin levels, leaf gas exchange, sugar content, and the expression of the *SWEET* gene family members.

## 2. Materials and Methods

### 2.1. AMF Inoculants

Six AMF species were tested: *Diversispora epigaea* (B.A. Daniels & Trappe) C. Walker & A. Schüßler (*De*), *Rhizophagus intraradices* (N.C. Schenck & G.S. Sm.) Sieverd., G.A. Silva & Oehl (*Ri*), *Paraglomus occultum* (C. Walker) J.B. Morton & D. Redecker (*Po*), *Entrophospora etunicata* (W.N. Becker & Gerd.) Błaszk., Niezgoda, B.T. Goto & Magurno (*Ee*), *Glomus heterosporum* G.S. Sm. & N.C. Schenck (*Gh*), and *Funneliformis mosseae* (T.H. Nicolson & Gerd.) C. Walker & A. Schüßler (*Fm*). All AMF strains were provided by the Institute of Root Biology, Yangtze University. These strains were propagated for 70 days using white clover (*Trifolium repens*) as a host plant under potted conditions. After removing aboveground biomass, colonized root fragments and growth substrates were collected as AMF inoculants, in which each gram of inoculum contained approximately 20 spores.

### 2.2. Plant Culture

The maize cultivar 'Zhengdan 958' was provided by the Food Crops Research Institute of the Henan Academy of Agricultural Sciences. The pot experiment was conducted in a greenhouse at the West Campus of Yangtze University from 15 July to 15 September, 2024. Pots (2.4 L) with dimensions of 16.5 cm top diameter, 14.5 cm bottom diameter, and 12.5 cm depth were used. Six AM fungal treatments were established: *De*, *Ri*, *Po*, *Ee*, *Gh*, and *Fm*. Inoculation treatments were carried out when maize was sown. Each treatment pot received 400 g of respective inoculum containing approximately 200 spores per 10 g, which was thoroughly mixed with 800 g of autoclaved growth substrate comprising soil and sand (3:1, *v*/*v*). Control (CK) pots were filled with 800 g of autoclaved substrate and 400 g of autoclaved inoculums plus of 3 mL of 20-μm-passed inoculum filters.

Prior to sowing, maize seeds were soaked in distilled water for 24 h. Four seeds were sown per pot. After seedling emergence, plants were thinned to two uniform seedlings per pot. Throughout the maize growth period, daily irrigation was applied to maintain well-watered moisture levels. The experiment used a completely randomized design with seven treatments (CK, *De*, *Ri*, *Po*, *Ee*, *Gh*, and *Fm*) and four replicates (pots) per treatment, totaling 28 pots. All the plants were grown in an environmentally controlled growth chamber, with the specific environmental conditions detailed in Liang et al. [32].

Pots were randomly arranged in the chamber to minimize positional effects (e.g., light/temperature gradients).

### 2.3. Determination of Plant Growth, Leaf Gas Exchange, and Root Mycorrhizal Colonization Rate

On the day of harvest, maize plant height was measured using a measuring tape, and stem diameter was determined with Vernier calipers. Subsequently, shoot and root biomass was dried at 80 °C for 48 h and then weighed. Prior to harvest, leaf gas exchange parameters were measured on the 3rd fully expanded leaves from the apex of two representative plants per treatment, with one leaf measured per plant. Between 9:30 and 11:30 a.m. on clear days, net photosynthetic rate (Pn), transpiration rate (Tr), and intercellular $CO_2$ concentration (Ci) were determined using a Yaxin-1102G portable photosynthesis system. The system was operated under the following conditions: photosynthetic photon flux density of 1000 μmol/m$^2$/s provided by an integrated red/blue LED light source, ambient $CO_2$ concentration (400 ± 10 μmol/mol), and a constant flow rate of 500 μmol/s. Leaf chamber temperature was maintained at ambient levels (25 °C) with relative humidity of 56%. Prior to data collection, each leaf was allowed to equilibrate for 60 s, followed by three consecutive measurements taken at 30 s intervals, with the average value recorded.

Harvested roots were thoroughly rinsed with tap water and cut into approximately 1.5 cm segments. Root colonization assessment was performed using trypan blue staining [33]. Root mycorrhizal colonization was examined under a microscope, and the root colonization rate was calculated as number of colonized root segments against total observed root segments.

### 2.4. Determinations of GRSP Levels

Following harvest, the growth substrate adhering to the root surfaces was gently shaken off and collected for the determination of GRSP levels. The contents of easily extractable GRSP (EEG) and difficultly extractable GRSP (DEG) were measured according to the method described by Wu et al. [34]. EEG was extracted by autoclaving 1.0 g soil in 20 mM citrate (pH 7.0) at 121 °C for 30 min, then centrifuging at 10,000× *g* for 3 min. The residues were subsequently extracted with 50 mM citrate (pH 8.0) for 60 min and centrifuged at 10,000× *g* for 3 min for DEG isolation, with both fractions analyzed using Bradford method with bovine serum albumin standards. Total GRSP (TG) content was calculated as the sum of EEG and DEG.

### 2.5. Determinations of Leaf Sugar Levels

Leaf sugar levels were assayed as per the protocol of Wen et al. [35]. Leaf samples (50 mg dry weight, dried at 65 °C and sieved at 0.5 mm) underwent dual extraction with 4 mL 80% ethanol at 80 °C for 40 min followed by centrifugation at 2500× *g*/min for 5 min. Combined supernatants were decolorized with 10 mg activated charcoal at 80 °C for 30 min and filtered. Sucrose quantification used resorcinol after alkaline hydrolysis: 150 μL extract + 150 μL 2 M NaOH (100 °C, 5 min), cooled, mixed with 2.1 mL 10 M HCl and 0.6 mL 0.1% ethanolic resorcinol, and incubated (80 °C, 10 min). The absorbance was measured at 480 nm against sucrose standards. Fructose quantification employed direct resorcinol reaction: 150 μL extract + 2.8 mL 10 M HCl + 0.8 mL 0.1% resorcinol (80 °C, 10 min). The absorbance was read at 480 nm using fructose standards. Glucose quantification followed enzymatic oxidation: 500 μL extract + 1 mL pre-warmed (30 °C, 2 min) enzyme reagent (glucose oxidase/peroxidase with *o*-dianisidine in acetate buffer, pH 5.5) and incubated (30 °C, 5 min). The reaction was terminated with 2 mL 10 M $H_2SO_4$, and the absorbance read at 460 nm against glucose standards.

### 2.6. *ZmSWEET* Gene Expression Analysis

Total RNA in leaf samples was isolated using the FastPure® Plant Total RNA Isolation Kit (Vazyme Biotech Co., Ltd., Nanjing, China), with RNA integrity verified through 1.2% agarose gel electrophoresis and concentration quantified via NanoDrop™ 2000 spectrophotometer. The RNA was reverse-transcribed into first-strand cDNA using the HiScript® II 1st Strand cDNA Synthesis Kit (Vazyme Biotech Co., Ltd., Nanjing, China), including a no-template control and a no-reverse-transcriptase control to monitor genomic DNA contamination and reagent purity. Quantitative real-time PCR (qRT-PCR) was performed using a CFX96 Touch™ system (Bio-Rad, Hercules, CA, USA) with Taq Pro Universal SYBR qPCR Master Mix (Vazyme Biotech Co., Ltd., Nanjing, China) in 20 μL reactions containing 50 ng cDNA and 0.4 μM gene-specific primers. Each qRT-PCR run incorporated no-template controls and no-amplification controls to detect potential contamination or nonspecific amplification. Expression profiles of 24 *ZmSWEET* homologs [36] were analyzed using primers designed via PrimerQuest™ Tool (Supplementary Material Table S1). The reference gene *ZmTubulin* was simultaneously amplified for normalization. Relative gene expression was calculated using the $2^{-\Delta\Delta Ct}$ method [37], with data $\log_2$-transformed for statistical analysis, after validation of amplification specificity through single-peak melt curves and confirmation of primer efficiencies via standard curve dilution series.

### 2.7. Statistical Analysis

Data processing was performed using Microsoft Excel 2016. Prior to statistical analysis, all datasets were tested for normality (Shapiro-Wilt test) and homogeneity of variance (Levene's test). If data violate these assumptions, logarithmic transformations were applied. Statistical significance was assessed through one-way analysis of variance (ANOVA) in IBM SPSS Statistics 27 (IBM, Armonk, NY, USA), followed by Fisher's Least Significant Difference (LSD) post hoc test for pairwise comparisons ($p < 0.05$). Experimental results were visualized using GraphPad Prism 8 (GraphPad, San Diego, CA, USA) and Origin 2024 software (OriginLab Corporation, Northampton, MA, USA).

## 3. Results

### 3.1. Effects of Different AMF Species on Maize Root Colonization

Root colonization analysis revealed significant variation in mycorrhizal colonization rates among AMF treatments. Characteristic symbiotic structures–including vesicles, hyphae, and arbuscules–were observed in all AMF-inoculated maize roots (Figure 1b–g), confirming successful colonization, while non-inoculated controls (CK) showed no colonization (Figure 1a). Colonization rates exhibited species-specific patterns: *Ri* and *Fm* demonstrated the highest colonization efficiencies at 66.88% and 68.33%, respectively, followed by *De* at 34.83% (Figure 1h). In contrast, significantly lower colonization rates were observed in *Po* (20.43%), *Ee* (21.30%), and *Gh* (25.50%), with a significant difference between *Po* and *Gh* inoculations.

### 3.2. Effects of Different AMF Species on Maize Growth

Maize growth responses to AMF inoculation exhibited significant variation across different fungal species (Figure 2a). Inoculation with *Ri*, *Po*, *Ee*, or *Fm* did not significantly affect maize plant height compared to CK (Figure 2b). However, maize inoculated with *De* exhibited 21.31% significantly greater plant height than CK. Among all treatments, *De*-inoculated plants achieved the maximum height (37.0 cm), with *De* treatments yielding significantly taller (12.03−33.31%) plants than those inoculated with *Po*, *Ee*, *Gh*, or *Fm*. For stem diameter, all six AMF-inoculated treatments showed significantly increased values relative to CK (Figure 2c). Notably, *Ee* inoculation produced the thickest stems, surpassing *De*, *Po*, *Gh*, and *Fm* by 31.47%, 29.89%, 17.21%, and 23.37%, respectively. Inoculation with AMF significantly enhanced both shoot and root biomass in maize plants relative to non-inoculated controls (Figure 2d,e). Shoot biomass exhibited significant increases in *De* (22.75%), *Ri* (51.82%), *Po* (36.94%), and *Ee* (32.47%) (Figure 2d), while root biomass showed a significant increase of *De* (32.39%), *Ri* (198.51%), *Po* (130.33%), *Ee* (64.75%), *Gh* (122.71%), and *Fm* (125.17%) (Figure 2e).

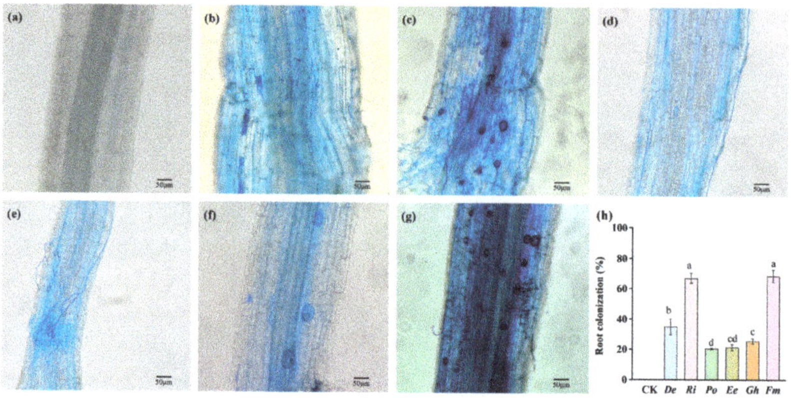


**Figure 1.** The root mycorrhizal colonization of maize (**a–g**) and the changes in root mycorrhizal colonization rate in maize colonized by six AMF (**h**). Data (means ± SD, $n = 4$) showed significant ($p < 0.05$; LSD tests) differences represented by different letters above the bars. (**a**): CK, inoculation without arbuscular mycorrhizal fungi; (**b**): *De*, *Diversispora epigaea*; (**c**): *Ri*, *Rhizophagus intraradices*; (**d**): *Po*, *Paraglomus occultum*; (**e**): *Ee*, *Entrophospora etunicata*; (**f**): *Gh*, *Glomus heterosporum*; (**g**): *Fm*, *Funneliformis mosseae*.

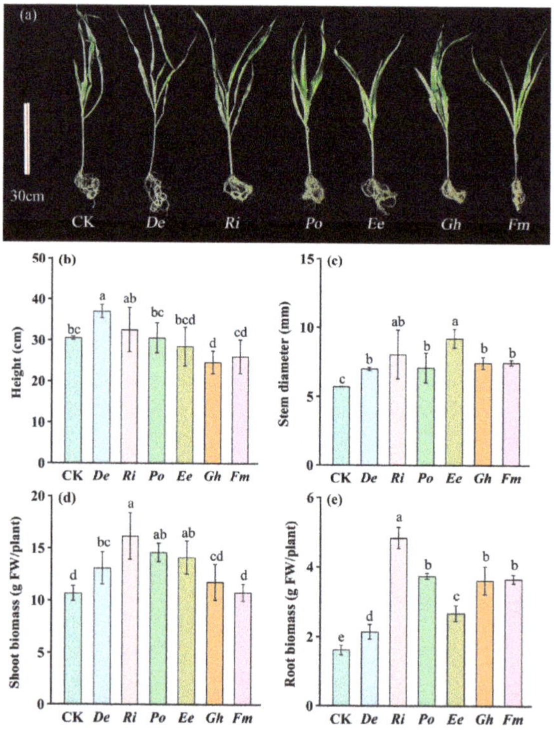


**Figure 2.** Plant growth behavior of different treated maize (**a**) and effects of six arbuscular mycorrhizal fungi on height (**b**), stem diameter (**c**), shoot biomass (**d**), and root biomass (**e**) of maize. Data (means ± SD, $n = 4$) showed significant ($p < 0.05$; LSD tests) differences represented by different letters above the bars. Abbreviation: CK, inoculation without arbuscular mycorrhizal fungi; *De*, *Diversispora epigaea*; *Ri*, *Rhizophagus intraradices*; *Po*, *Paraglomus occultum*; *Ee*, *Entrophospora etunicata*; *Gh*, *Glomus heterosporum*; *Fm*, *Funneliformis mosseae*.

### 3.3. Effects of Different AMF Species on Leaf Gas Exchange Parameters

All AMF inoculations enhanced photosynthetic performance to a certain extent relative to CK, though species-specific variations were prominent (Figure 3a–c). *Ri, Ee,* and *Fm* significantly increased Pn by 39.24%, 91.73%, and 83.67%, respectively, while *De, Po,* and *Gh* showed no significant differences (Figure 3a). Tr increased significantly with *Ri* (80.78%), *Ee* (43.59%), and *Fm* (62.03%) inoculations, whereas *De* and *Gh* had no notable impact (Figure 3b). *De, Ri, Po,* and *Gh* elevated Ci levels by 53.18%, 40.04%, 39.01%, and 43.12%, while *Ee* and *Fm* treatments remained statistically unchanged (Figure 3c).

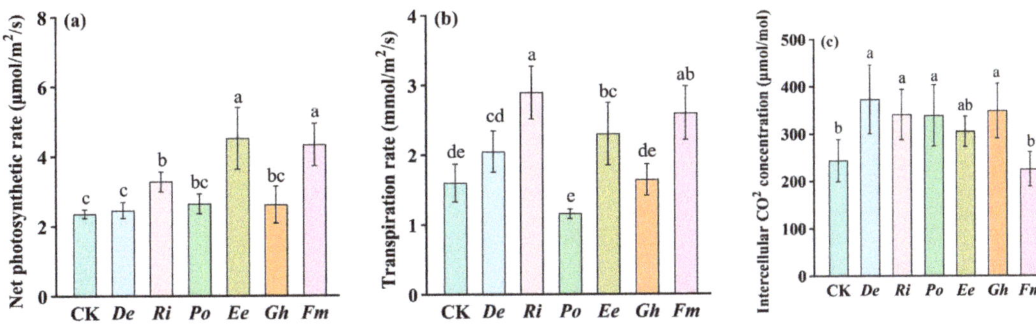


**Figure 3.** Effects of six arbuscular mycorrhizal fungi on leaf net photosynthetic rate (**a**), transpiration rate (**b**), and intercellular $CO_2$ concentration (**c**) of maize. Data (means ± SD, *n* = 4) showed significant ($p < 0.05$; LSD tests) differences represented by different letters above the bars. Refer to Figure 2 for the abbreviations.

### 3.4. Effects of Different AMF Species on GRSP Levels of Substrate Soil

AMF inoculation significantly modulated GRSP accumulation in the maize substrate soil compared to CK (Figure 4a–c). Significant increases in EEG occurred with *De* (9.55%), *Ri* (11.04%), *Po* (14.02%), *Ee* (5.97%), and *Gh* (11.63%), compared to CK. *Fm* showed a non-significant reduction in EEG. In addition, *De, Ri,* and *Po* significantly enhanced DEG levels by 3.10%, 4.53%, and 6.56%, respectively, while *Ee* markedly reduced DEG by 9.97%, compared to CK. *Gh* and *Fm* exhibited non-significant alterations in DEG relative to CK. TG was highest in *De, Ri,* and *Po* inoculated substrate soils, significantly surpassing levels in *Ee*, CK, and *Fm* treatments.

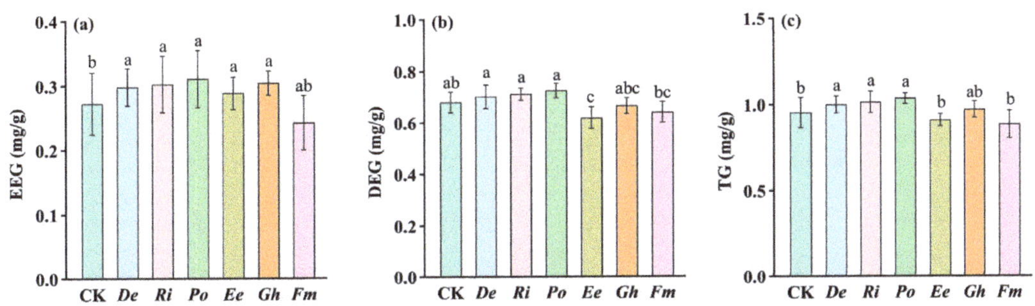


**Figure 4.** Effects of six arbuscular mycorrhizal fungi on substrate soil EEG (**a**), DEG (**b**), and TG (**c**) levels of maize. Data (means ± SD, *n* = 4) showed significant ($p < 0.05$; LSD tests) differences represented by different letters above the bars. Abbreviation: EEG, easily extractable glomalin-related soil protein; DEG, difficultly extractable glomalin-related soil protein; TG, total glomalin-related soil protein. Refer to Figure 2 for the other abbreviations.

### 3.5. Effects of Different AMF Species on Leaf Sucrose, Fructose, and Glucose Concentrations

Inoculation with different AMF species had varying effects on sugar component concentrations in maize leaves (Figure 5a–c). Compared with CK, sucrose concentrations significantly increased in the *De* (326.86%), *Ri* (93.04%), *Po* (544.50%), *Ee* (132.69%), *Gh* (179.61%), and *Fm* (184.47%) treatments. Glucose concentrations significantly increased in the *De* (369.85%), *Po* (709.74%), *Ee* (191.95%), *Gh* (471.91%), and *Fm* (524.34%) treatments. The *Ri* treatment showed no significant increase in glucose concentrations, compared with CK. Compared with CK, fructose concentrations significantly increased in the *Po* (148.20%) and *Fm* (77.62%) treatments, while the *De*, *Ri*, *Ee*, and *Gh* treatments had no significant effect on fructose concentrations (Figure 5c). Additionally, *Po* treatment exhibited the greatest leaf sucrose, glucose, and fructose concentrations compared with the other inoculations and CK.

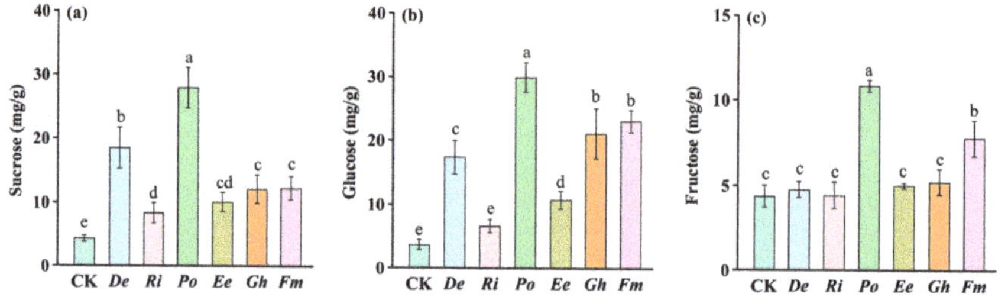


**Figure 5.** Effects of six arbuscular mycorrhizal fungi on leaf sucrose (**a**), glucose (**b**), and fructose (**c**) levels of maize. Data (means ± SD, $n = 4$) showed significant ($p < 0.05$; LSD tests) differences represented by different letters above the bars. Refer to Figure 2 for the abbreviations.

### 3.6. Effects of Different AMF Species on the Expression of Leaf ZmSWEETs

Following the removal of genes *ZmSWEET6a*, *ZmSWEET15a*, and *ZmSWEET17a* (which showed no detectable signals across all treatments), the expression levels of the remaining 21 genes were $log_2$-transformed for standardization. The expression patterns of *ZmSWEET* genes varied significantly under different AMF treatments (Figure 6). Specifically, *ZmSWEET1b*, *ZmSWEET3a*, *ZmSWEET3b*, *ZmSWEET4a*, *ZmSWEET4b*, *ZmSWEET14a*, and *ZmSWEET16* were consistently upregulated in AMF-inoculated maize leaves. Conversely, *ZmSWEET6b*, *ZmSWEET11b*, *ZmSWEET12a*, *ZmSWEET13a*, *ZmSWEET13b*, *ZmSWEET13c*, and *ZmSWEET17b* exhibited consistent downregulation under mycorrhization conditions. The remaining genes showed either up- or downregulation patterns depending on the specific AMF treatment applied.

### 3.7. Results Analysis of the PCA

The principal component analysis (PCA) showed the relationships between 21 *ZmSWEET* gene expressions, sugar (sucrose, fructose, and glucose), root AMF colonization, and different experimental groups (Figure 7). PC1 (42.87%) and PC2 (19.65%) collectively account for 62.52% of the total variance. Samples form separate three clusters, where CK and *Fm* each formed separate clusters, while *De*, *Ee*, *Gh*, *Ri*, and *Po* grouped together. It suggested different treatments (or microbial inoculations) induce unique responses in *ZmSWEET* expression and carbon metabolism. Sucrose, glucose, and fructose strongly correlate with PC1 (positive axis), clustering with *ZmSWEET4a*, *ZmSWEET15b*, *ZmSWEET14b*, etc. This suggested these genes were closely linked to sugar accumulation under *Po*, *Gh*, *Ee* and *Fm* treatments. Root colonization associated with PC1 (positive axis), clustering with sugar-related *ZmSWEETs* (such as *ZmSWEET1b*, *ZmSWEET14a* and *ZmSWEETb-2*) and sugars. Correlation analysis further revealed that the root AMF colonization rate was significantly positively correlated with *ZmSWEET3a* ($r = 0.55$, $p < 0.01$), *ZmSWEET3b* ($r = 0.51$, $p < 0.01$), *ZmSWEET4c* ($r = 0.58$, $p < 0.01$), and *ZmSWEET14a* ($r = 0.48$, $p < 0.01$), but significantly negatively correlated with *ZmSWEET1a* ($r = -0.64$, $p < 0.01$), *ZmSWEET11a* ($r = -0.46$, $p < 0.05$), *ZmSWEET11b* ($r = -0.50$, $p < 0.01$), *ZmSWEET12a* ($r = -0.60$, $p < 0.01$), *ZmSWEET12b* ($r = -0.59$, $p < 0.01$), *ZmSWEET13a* ($r = -0.73$, $p < 0.01$), *ZmSWEET13b* ($r = -0.82$, $p < 0.01$), *ZmSWEET13c* ($r = -0.79$, $p < 0.01$), and *ZmSWEET17b* ($r = -0.82$, $p < 0.01$). This hinted that root AMF colonization may directly/indirectly regulate specific *SWEET*-mediated sugar transport/accumulation. Groups like *Gh*, *Po*, and *Ee* clustered near sugars, suggesting these fungi enhanced *SWEET*-driven sugar uptake/transport in roots. *ZmSWEET4a*, *ZmSWEET4b*, *ZmSWEET14b ZmSWEET15b*, and *ZmSWEET16* emerged as central to sugar metabolism under symbiotic conditions. Sucrose and glucose were pivotal, driving PC1 variability and linking mycorrhizal colonization to *SWEET* expression.

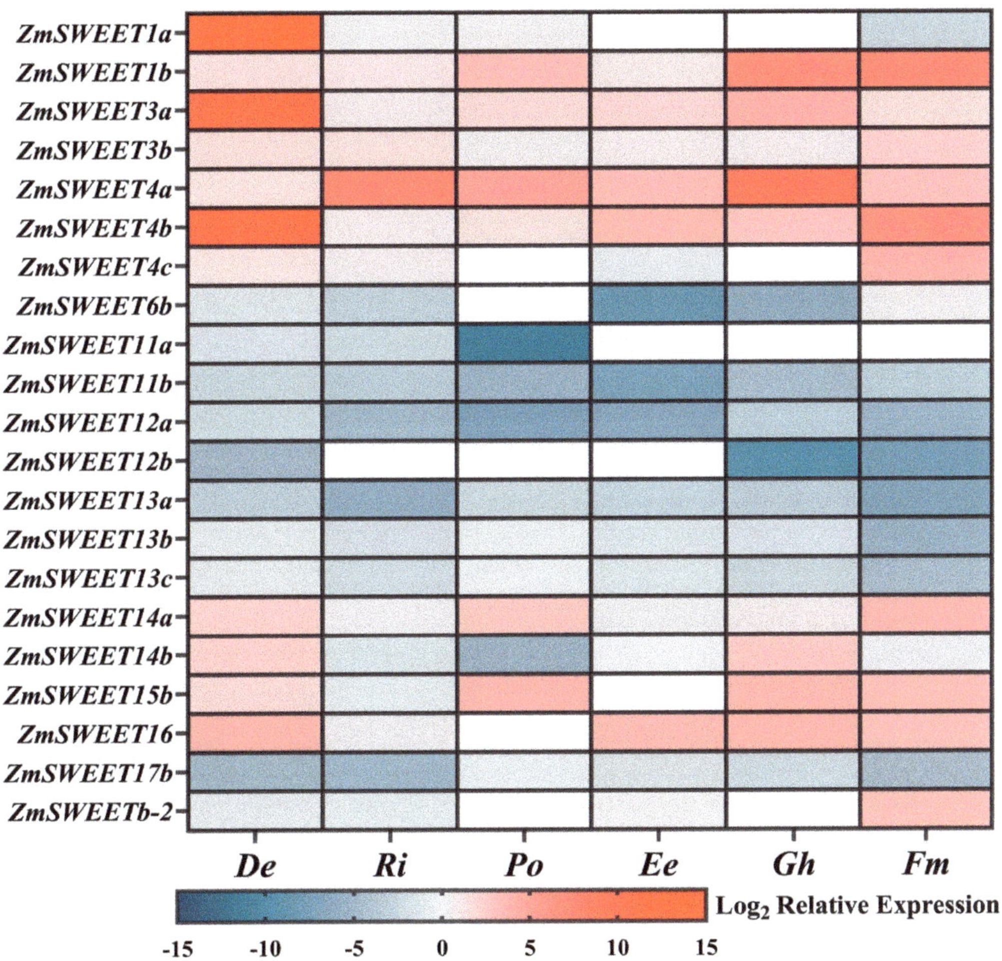


**Figure 6.** Effects of six arbuscular mycorrhizal fungi on the expression of leaf *ZmSWEETs* of maize. White indicates undetected the gene expression. Refer to Figure 2 for the abbreviations.

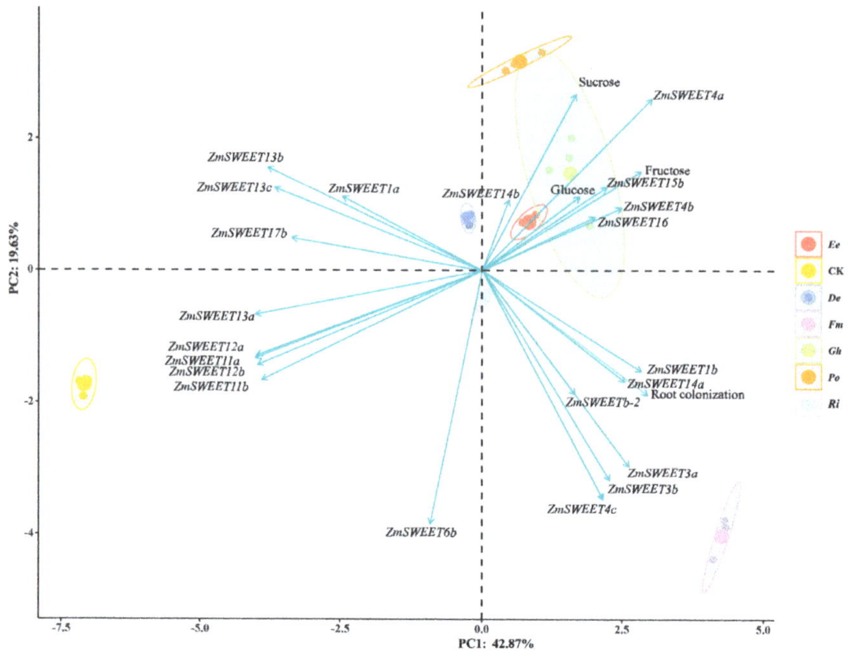


**Figure 7.** Principal component analysis of root mycorrhizal colonization, sugars, and *ZmSWEETs* in maize under different AMF inoculations.

## 4. Discussion

AMF can colonize plant roots to form symbiotic associations that regulate plant growth, while the magnitude of mycorrhizal effects is influenced by multiple factors, including AMF species, soil environment, host plant type, and human activities [38]. Mycorrhizal colonization rate serves as a critical indicator of AMF expansion capacity within roots and reflects the symbiotic compatibility between the fungi and host plants [39,40]. In our study, all inoculated AMF strains successfully established symbiosis with maize roots, with *Ri* and *Fm* achieving peak colonization (66.88% and 68.33%) and *Po* exhibiting the lowest colonization (20.43%). These results demonstrate distinct host-symbiont compatibility hierarchies, identifying *Ri* and *Fm* as the most effective colonizers of maize roots under experimental conditions. This reflected that strain-dependent host recognition in AMF-maize systems [41]. The superior colonization rates of *Ri* and *Fm* likely result from strain-specific adaptations that enhance symbiotic compatibility including host recognition efficiency, cell invasion strategies, and nutrient-exchange synergy, which must be further studied. *Ri* and *Fm* achieved high colonization rates, whereas their functional performance (such as growth promotion) were not consistent, dependent on the combination of the plant and fungal genotypes and on environmental conditions [38,42].

Our results showed that AMF inoculation could significantly increase plant biomass, indicating that AMF could promote the absorption of nutrients by plants and increase plant biomass. *Ri*, *Po*, *Ee*, and *Fm* did not significantly alter plant height compared to CK, while *De* uniquely promoted plant height, yielding the tallest plants among all treatments. The differential influence of AMF species on maize growth reveals a complex interplay between symbiotic efficiency and functional specialization. This height advantage was particularly pronounced relative to *Po*, *Ee*, *Gh*, and *Fm* inoculations, suggesting that *De* may enhance auxin signaling or improve photosynthate allocation toward stem elongation [43,44]. In contrast, all AMF-inoculated plants exhibited significantly thicker stems than CK, underscoring a universal benefit of mycorrhization on radial development. Remarkably, *Ee*—despite its low colonization rate (21.30%)—produced the stoutest stems, exceeding other AMF treatments. This implies distinct physiological mechanisms: *Ee* may optimize carbon partitioning to stem vasculature or enhance water/nutrient flux through specialized hyphal networks. The stem response divergence highlights that colonization extent alone cannot predict functional outcomes; *Gh* showed higher colonization than *Ee* yet lower stem enhancement, indicating strain-specific functional hierarchies.

Photosynthetic capacity serves as a critical physiological foundation for plant growth and yield formation [45]. Singh et al. pointed out that inoculation with AMF could enhance photosynthetic capacity in leaves of host plants [46]. In our study, *Ri*, *Ee*, and *Fm* drove substantial increases in Pn (39.24–91.73%), reflecting superior efficiency in facilitating Rubisco activation and electron transport chain coordination [47]. The concurrent Tr elevation in these treatments (43.59–80.78%) aligns with AMF-mediated improvement in stomatal conductance and hydraulic conductivity, a well-documented mechanism for enhancing $CO_2$ diffusion into mesophyll cells [48]. Notably, *Ee*'s exceptional Pn boost (91.73%)—despite its low root colonization rate (21.30%)—suggests that certain AMF strains prioritize the modulation of chloroplast development or photophosphorylation efficiency [49]. Conversely, the lack of significant Pn and Tr response in *De* and *Gh* indicates functional trade-offs; these strains may allocate resources toward growth structural investment (e.g., stem thickening) rather than photosynthetic optimization, as evidenced by *De*'s height promotion and *Gh*'s stem diameter neutrality. The observed 39.01–53.18% significant increase in Ci under *De*, *Ri*, *Po*, and *Gh* treatments could reflect a limitation in carboxylation or stomatal conductance, which warrants in-depth study.

GRSP, which is a metal-ion-containing glycoprotein produced by AMF, could enter the soil through the mycelium and spores of AMF. Subsequently, it will bind the micro-aggregates together, while also increasing the organic carbon content in the soil and enhancing the structural stability of the soil [34,50]. Our study revealed that the EEG elevation across *De*, *Ri*, *Po*, *Ee*, and *Gh* treatments (5.97–14.02%) demonstrated a photosynthate-derived carbon toward fresh glomalin production. However, the sharp divergence in DEG responses—where *De*, *Ri*, and *Po* increased DEG (3.10–6.56%) while *Ee* reduced it markedly—suggests distinct glomalin stabilization pathways among AMF taxa. Notably, the triad of *De*, *Ri*, and *Po* emerged as optimal GRSP producers, driving parallel increases in EEG and TG. Conversely, *Ee*'s strategy—elevating EEG while suppressing DEG—may represent an ecological adaptation. The non-responsiveness of *Fm* across GRSP fractions reinforces its earlier observed metabolic economy; despite high root colonization, it minimizes glomalin production, possibly redirecting carbon toward hyphal network expansion. This GRSP differentiation underscores that AMF functional traits extend beyond plant growth modulation to direct soil engineering—a critical consideration when selecting strains for ecosystem restoration versus crop productivity [51].

Sugar is a secondary messenger and an important organic permeate in plants. AMF could obtain hexose, mainly glucose, which is one of the most important categories of sugar, by the hydrolysis of sucrose from host plants. They could also convert these sugars into trehalose and glycogen, which are typical fungal carbohydrates [26,52]. Hyphal carbon demand alters source leaf sugar equilibrium [53]. In our study, a significant increase in sucrose content (93.04–544.50%) was observed in all AMF treatment groups, indicating an enhanced load capacity of the phloem and an improved coordination between the source and sink. Similar results were also reported by Wu et al. [52]. However, the extreme sucrose accumulation in *Po*-inoculated leaves (544.50%)—coupled with its parallel fructose surge (709.74%)—suggests exceptional carbon fixation efficiency. The elevated sucrose levels could potentially support increased growth and development by providing more carbon resources for processes such as root development, reproductive growth, or defense mechanisms [54]. Fructose responses exhibited greater AMF specificity: *Po* and *Fm* drove substantial increases, while *De, Ri, Ee,* and *Gh* exhibited no significant change, reflecting differential fructose modulation. The absence of fructose response in *De, Ri, Ee,* and *Gh* despite its sucrose elevation indicates preferential carbon channeling toward root allocation rather than hexose accumulation. The dynamic changes in glucose levels further highlighted the functional differences, for *Po* and *Fm* significantly increased the glucose content (77.62–148.20%), indicating the existence of glucose-induced effects specific to different strains. The change in fructose and glucose is due to the accelerated cleavage of sucrose into glucose and fructose by mycorrhization [55]. In our study, sucrose, glucose and fructose have a high correlation with the root colonization parameters in the PC1 cluster analysis, confirming that AMF symbionts actively adjust the carbon allocation pathways. However, more research needs to focus on how much sugars are distributed from the leaves to the roots and the root mycorrhizae.

It is known that the *ZmSWEET* family is phylogenetically divided into four clades. Clades I (*SWEETs1–3*) and II (*SWEETs4–8*) contain genes with glucose transport activity, Clade III (*SWEETs9–15*) genes mediate sucrose transport, and Clade IV (*SWEETs16–17*) includes fructose transport [56,57]. The consistent upregulation of *ZmSWEET1b/3a/3b/4a/4b/14a* and *16* across all AMF treatments reveals a core transcriptional reprogramming essential for mycorrhizal symbiosis. The conserved response may explain the evolutionary conservation among these *ZmSWEETs*, functional specialization, and host-fungal coordination. These genes likely mediate symbiotic carbon allocation by facilitating sucrose efflux toward fungal interfaces, as *SWEET* transporters are established regulators of apoplastic sugar transfer [58]. Notably, *ZmSWEET4a/b* homologs in rice (*OsSWEET11/15*) similarly exhibited AMF-inducible expression, suggesting conserved roles in delivering carbon rewards to fungal partners [59]. *SWEET4* homologs in other species mediate sucrose efflux to support microbial symbionts (e.g., arbuscular mycorrhizae or rhizobia) [60]. In maize, the upregulation of the *ZmSWEET* gene indicates that it can facilitate the transportation of more sugar to the AMF, thereby promoting the growth of the AMF and the exchange of nutrients. The association of *ZmSWEET4a/15b* and *14b* expression with sugars in PCA suggests that these transporters served as central metabolic hubs for photoassimilates toward mycorrhizal interfaces. This coupling implied that AMF colonization directly activates *SWEET* mediated sucrose efflux into the apoplast, where fungal hyphae access these carbon resources [61]. Some *SWEET* genes may have been recruited for mycorrhizal symbiosis in the early stage of plant evolution [28].

The localization results processed by *Ri* and *Gh* indicated that both of these two AMF strains exhibited distinct mycelial and sugar cluster growth characteristics. This suggests that these two AMF species perform exceptionally well in jointly establishing absorption capabilities. They could promote the growth of mycelia in the roots, then simultaneously upregulated the key *ZmSWEET* type proteins to facilitate carbon transfer. This aligned with sucrose accumulation in earlier observed of *Ri* treatment and EEG elevation in *Gh* treatment, revealing a unified strategy where fungal partners manipulate host transport machinery to optimize resource exchange. The dominant role of sucrose in driving the PC1 variation highlights its role as the main sugar species in leaves responsible for long-distance transport in the phloem of plants to roots and cleave glucose for root mycorrhizas [62]. Conversely, the uniform downregulation of *ZmSWEET6b/11b/12a/13a-c* and *17b* indicated suppression of alternative carbon sinks—potentially redirecting photoassimilates away from non-symbiotic tissues to fuel mycorrhizal networks. Correlation analysis also presented a significantly negative correlation between root AMF colonization rates with

*ZmSWEET11b*, *ZmSWEET12a*, *ZmSWEET13a-c*, and *ZmSWEET17b*. This transcriptional dichotomy aligns with the "symbiotic trade-off" model, where plants prioritize carbon investment into AMF-colonized roots over endogenous metabolic pathways [62]. The strain-dependent expression of remaining *ZmSWEET* genes highlighted AMF-specific modulation of host sugar transporters [63]. Their co-regulation of *ZmSWEETs* with root colonization further supports a feedback loop where AMF-derived signals (e.g., Myc-LCOs) transcriptionally reprogram sugar transport to sustain symbiosis [64]. Based on these, we can speculate that the inoculation of AMF reprogrammed *ZmSWEET* expression in maize by a species-specific manner, with core genes mediating carbon allocation to support symbiosis. This specific mechanism and which are core genes still requires further in-depth study.

## 5. Conclusions

In short, the inoculation of AMF had a positive promoting effect on the growth and leaf photosynthetic capacity of maize, and also had a positive promoting effect on the substrate GRSP levels. More importantly, AMF could affect the expression of the *ZmSWEET* sugar transporter genes in maize by consistently upregulating *ZmSWEET1b/3a/3b/4a/4b/14a/16*, and downregulating *ZmSWEET6b/11b/12a/13a/13b/13c/17b*. The colonization of AMF and sugar accumulation was closely related to the expression of specific *ZmSWEET* genes. High colonization reflects compatibility, but not necessarily functionality. Strain selection for agriculture should prioritize performance metrics (yield and stress resilience) over colonization rates alone. These findings disclosed that AMF mediates the accumulation of leaf sugar in maize through species-specific reprogramming of the expression of *SWEET* sugar transporter genes, so as to support the establishment and maintenance of the symbiotic relationship. Our study laid a foundation for further exploration of the signal pathways involved in the interaction of AMF and *SWEET* genes.

**Supplementary Materials:** The following supporting information can be downloaded at: https://www.mdpi.com/article/10.3390/agriculture15161790/s1, Table S1: Specific primer sequences of selected genes in qRT-PCR.

**Author Contributions:** Conceptualization, Q.-S.W.; Methodology, G.-X.H.; Material preparation, G.-X.H. and C.G.; Data determination and statistical analysis, G.-X.H., F.-L.Z. and C.G.; Supervision, Q.-S.W. and C.G.; Writing—original draft preparation, G.-X.H.; Writing—review and editing, C.G., X.-B.G., F.-L.Z., Y.-N.Z. and Q.-S.W. All authors have read and agreed to the published version of the manuscript.

**Funding:** This work was supported by the National Natural Science Foundation of China (31960616), the Construction of Modern Agriculture (tea) Industry Technology System (CARS-19) (China), Guizhou Key Laboratory of Agricultural Microbiology Construction Project (Qiankehe-pingtai[2025]029), Tea Science and Technology Experimental Demonstration Base Construction in Sinan County for 2025 ZSYS (Guizhou Province), and Research and Development of Special Microbial Fertilizer for Tea Seedling Cultivation Using Arbuscular Mycorrhizal Fungi [2025520101000147].

**Institutional Review Board Statement:** Not applicable.

**Informed Consent Statement:** Not applicable.

**Data Availability Statement:** All the data supporting the findings of this study are included in this article.

**Acknowledgments:** The authors gratefully acknowledge support from as shown under Funding. Special thanks support from Guizhou Yibai Billion Biotechnology Co., Ltd., China (2025520101000147).

**Conflicts of Interest:** The authors declare no conflicts of interest.

## References


1. Wang, J.J.; Yan, D.; Liu, R.; Wang, T.; Lian, Y.J.; Lu, Z.; Hong, Y.; Wang, Y.; Li, R. The physiological and molecular mechanisms of exogenous melatonin promote the seed germination of maize (*Zea mays* L.) under salt stress. *Plants* **2024**, *13*, 2142. [CrossRef] [PubMed]
2. Chaudhary, D.P.; Kumar, S.; Yadav, O.P. Nutritive value of maize: Improvements, applications and constraints. In *Maize: Nutrition Dynamics and Novel Uses*; Chaudhary, D.P., Kumar, S., Langyan, S., Eds.; Springer: Berlin/Heidelberg, Germany, 2013; pp. 3–17.
3. National Bureau of Statistics of China. National Grain Production Data for 2024. Available online: https://www.stats.gov.cn/ (accessed on 20 February 2025).
4. Borriello, R.; Lumini, E.; Girlanda, M.; Bonfante, P.; Bianciotto, V. Effects of different management practices on arbuscular mycorrhizal fungal diversity in maize fields by a molecular approach. *Biol. Fert. Soils* **2012**, *48*, 911–922. [CrossRef]
5. Chen, X.; Wu, X.L.; Liu, S.R.; Hu, X.C.; Liu, C.Y. Effects of AMF on photosynthetic characteristics and gene expressions of tea plants under drought stress. *Acta Hortic. Sin.* **2024**, *51*, 2358–2370.
6. Bunn, R.A.; Corrêa, A.; Joshi, J.; Kaiser, C.; Lekberg, Y.; Prescott, C.E.; Sala, A.; Karst, J. What determines transfer of carbon from plants to mycorrhizal fungi? *New Phytol.* **2024**, *244*, 1199–1215. [CrossRef]
7. Liu, Z.; Cheng, X.F.; Zou, Y.N.; Srivastava, A.K.; Alqahtani, M.D.; Wu, Q.S. Negotiating soil water deficit in mycorrhizal trifoliate orange plants: A gibberellin pathway. *Environ. Exp. Bot.* **2024**, *219*, 105658. [CrossRef]
8. Zou, Y.N.; Wu, Q.S.; Kuča, K. Unravelling the role of arbuscular mycorrhizal fungi in mitigating the oxidative burst of plants under drought stress. *Plant Biol.* **2021**, *23*, 50–57. [CrossRef]
9. Ma, J.Q.; Wang, W.Q.; Yang, J.; Qin, S.F.; Yang, Y.S.; Sun, C.; Pei, G.; Zeeshan, M.; Liao, H.; Liu, L. Mycorrhizal symbiosis promotes the nutrient content accumulation and affects the root exudates in maize. *BMC Plant Biol.* **2022**, *22*, 64. [CrossRef]
10. Zhou, J.; Zhang, L.; Feng, G.; George, T.S. Arbuscular mycorrhizal fungi have a greater role than root hairs of maize for priming the rhizosphere microbial community and enhancing rhizosphere organic P mineralization. *Soil Biol. Biochem.* **2022**, *171*, 108713. [CrossRef]
11. Wright, S.F.; Upadhyaya, A. A survey of soils for aggregate stability and glomalin, a glycoprotein produced by hyphae of arbuscular mycorrhizal fungi. *Plant Soil* **1998**, *198*, 97–107. [CrossRef]
12. Cao, M.A.; Wang, P.; Hashem, A.; Wirth, S.; Abd_Allah, E.F.; Wu, Q.S. Field inoculation of arbuscular mycorrhizal fungi improves fruit quality and root physiological activity of citrus. *Agriculture* **2021**, *11*, 1297. [CrossRef]
13. Meng, L.L.; He, J.D.; Zou, Y.N.; Wu, Q.S.; Kuča, K. Mycorrhiza-released glomalin-related soil protein fractions contribute to soil total nitrogen in trifoliate orange. *Plant Soil Environ.* **2020**, *66*, 183–189. [CrossRef]
14. Zhao, H.; Liu, Z.; Han, Y.; Cao, J. Impact of silver nanoparticles on arbuscular mycorrhizal fungi and glomalin-related soil proteins in the rhizosphere of maize seedlings. *Diversity* **2024**, *16*, 273. [CrossRef]
15. Ramírez-Flores, M.R.; Bello-Bello, E.; Rellán-Álvarez, R.; Sawers, R.J.; Olalde-Portugal, V. Inoculation with the mycorrhizal fungus *Rhizophagus irregularis* modulates the relationship between root growth and nutrient content in maize (*Zea mays* ssp. *mays* L.). *Plant Direct* **2019**, *3*, e00192. [CrossRef]
16. Bisht, A.; Gupta, M.M. Arbuscular mycorrhiza fungi resources for sustainable and climate-smart cultivation of maize. In *Fungal Resources for Sustainable Economy: Current Status and Future Perspectives*; Singh, I., Rajpal, V.R., Navi, S.S., Eds.; Springer: Berlin/Heidelberg, Germany, 2023; pp. 299–317.
17. Aguégué, M.R.; Ahoyo Adjovi, N.R.; Agbodjato, N.A.; Noumavo, P.A.; Assogba, S.; Salami, H.; Salako, V.K.; Ramón, R.; Baba-Moussa, F.; Adjanohoun, A. Efficacy of native strains of arbuscular mycorrhizal fungi on maize productivity on ferralitic soil in Benin. *Agric. Res.* **2021**, *11*, 627–641. [CrossRef]
18. Assogba, S.; Noumavo, P.A.; Dagbenonbakin, G.; Agbodjato, N.; Akpode, C.; Koda, A.D.; Aguegue, R.M.; Bade, F.; Adjanohoun, A.; Rodriguez, A.F.; et al. Improvement of maize productivity (*Zea mays* L.) by mycorrhizal inoculation on ferruginous soil in center of Benin. *Int. J. Sustain. Agric. Res.* **2017**, *4*, 63–76. [CrossRef]
19. Wu, X.J.; Li, Z.F.; Guo, P.P.; Zhang, L. Effects of different arbuscular mycorrhizal fungi on the growth and the nitrogen and phosphorus absorption of sweet corn seedlings. *J. Trop. Biol.* **2023**, *14*, 167–172.
20. Koda, A.D.; Dagbenonbakin, G.; Assogba, F.; Noumavo, P.A.; Agbodjato, N.A.; Assogba, S.; Aguegue, R.M.; Adjanohoun, A.; Rivera, R.; de la Noval Pons, B.M.; et al. Maize (*Zea mays* L.) response to mycorrhizal fertilization on ferruginous soil of northern Benin. *J. Exp. Biol. Agric. Sci.* **2018**, *6*, 919–928. [CrossRef]
21. Wang, Y.J.; Wu, Q.S. Influence of sugar metabolism on the dialogue between arbuscular mycorrhizal fungi and plants. *Hortic. Adv.* **2023**, *1*, 2. [CrossRef]
22. Walmsley, A.R.; Barrett, M.P.; Bringaud, F.; Gould, G.W. Sugar transporters from bacteria, parasites and mammals: Structure-Activity relationships. *Trend. Biochem. Sci.* **1998**, *23*, 476–481. [CrossRef]
23. Julius, B.T.; Leach, K.A.; Tran, T.M.; Mertz, R.A.; Braun, D.M. Sugar transporters in plants: New insights and discoveries. *Plant Cell Physiol.* **2017**, *58*, 1442–1460. [CrossRef]

24. Chen, L.Q.; Hou, B.H.; Lalonde, S.; Takanaga, H.; Hartung, M.L.; Qu, X.Q.; Guo, W.J.; Kim, J.G.; Underwood, W.; Chaudhuri, B. Sugar transporters for intercellular exchange and nutrition of pathogens. *Nature* **2010**, *468*, 527–532. [CrossRef]
25. Ji, J.L.; Yang, L.M.; Fang, Z.Y.; Zhang, Y.Y.; Zhuang, M.; Lv, H.H.; Wang, Y. Plant SWEET family of sugar transporters: Structure, evolution and biological functions. *Biomolecules* **2022**, *12*, 205. [CrossRef]
26. Zheng, F.L.; Wang, Y.J.; Hashem, A.; Abd_Allah, E.F.; Wu, Q.S. Mycorrhizae with *Funneliformis mosseae* regulate the trehalose synthesis and sucrose cleavage for enhancing drought tolerance in trifoliate orange. *Sci. Hortic.* **2024**, *337*, 113486. [CrossRef]
27. Díaz, V.; Villalobos, M.; Arriaza, K.; Flores, K.; Hernández-Saravia, L.P.; Velásquez, A. Decoding the dialog between plants and arbuscular mycorrhizal fungi: A molecular genetic perspective. *Genes* **2025**, *16*, 143. [CrossRef] [PubMed]
28. Manck-Götzenberger, J.; Requena, N. Arbuscular mycorrhiza symbiosis induces a major transcriptional reprogramming of the potato *SWEET* sugar transporter family. *Front. Plant Sci.* **2016**, *7*, 487. [CrossRef] [PubMed]
29. Zheng, L.; Zhao, S.; Zhou, Y.; Yang, G.; Chen, A.; Li, X.; Wang, J.; Tian, J.; Liao, H.; Wang, X. The soybean sugar transporter *GmSWEET6* participates in sucrose transport towards fungi during arbuscular mycorrhizal symbiosis. *Plant Cell Environ.* **2024**, *47*, 1041–1052. [CrossRef]
30. An, J.; Zeng, T.; Ji, C.; de Graaf, S.; Zheng, Z.; Xiao, T.T.; Deng, X.; Xiao, S.; Bisseling, T.; Limpens, E. A *Medicago truncatula SWEET* transporter implicated in arbuscule maintenance during arbuscular mycorrhizal symbiosis. *New Phytol.* **2019**, *224*, 396–408. [CrossRef] [PubMed]
31. Graham, J.H.; Duncan, L.W.; Eissenstat, D.M. Carbohydrate allocation patterns in citrus genotypes as affected by phosphorus nutrition, mycorrhizal colonization and mycorrhizal dependency. *New Phytol.* **1997**, *135*, 335–343. [CrossRef]
32. Liang, S.M.; Hashem, A.; Abd_Allah, E.F.; Wu, Q.S. Transcriptomic analysis reveals potential roles of polyamine and proline metabolism in waterlogged peach roots inoculated with *Funneliformis mosseae* and *Serendipita indica*. *Tree Physiol.* **2025**, *45*, tpaf013. [CrossRef]
33. Phillips, J.M.; Hayman, D.S. Improved procedures for clearing roots and staining parasitic and vesicular-arbuscular mycorrhizal fungi for rapid assessment of infection. *Trans. Br. Mycol. Soc.* **1970**, *55*, 158–161. [CrossRef]
34. Wu, Q.S.; Li, Y.; Zou, Y.N.; He, X.H. Arbuscular mycorrhiza mediates glomalin-related soil protein production and soil enzyme activities in the rhizosphere of trifoliate orange grown under different P levels. *Mycorrhiza* **2015**, *25*, 121–130. [CrossRef]
35. Wen, Y.; Zhou, L.J.; Xu, Y.J.; Hashem, A.; Abd_Allah, E.F.; Wu, Q.S. Growth performance and osmolyte regulation of drought-stressed walnut plants are improved by mycorrhiza. *Agriculture* **2024**, *14*, 367. [CrossRef]
36. Zhu, J.L.; Zhou, L.; Li, T.F.; Ruan, Y.Y.; Zhang, A.; Dong, X.; Zhu, Y.; Li, C.; Fan, J. Genome-wide investigation and characterization of *SWEET* gene family with focus on their evolution and expression during hormone and abiotic stress response in maize. *Genes* **2022**, *13*, 1682. [CrossRef]
37. Livak, K.J.; Schmittgen, T.D. Analysis of relative gene expression data using real-time quantitative PCR and the $2^{-\Delta\Delta Ct}$ method. *Methods* **2001**, *25*, 402–408. [CrossRef]
38. Berger, F.; Gutjahr, C. Factors affecting plant responsiveness to arbuscular mycorrhiza. *Curr. Opin. Plant Biol.* **2021**, *59*, 101994. [CrossRef]
39. Gange, A.C.; Ayres, R.L. On the relation between arbuscular mycorrhizal colonization and plant 'benefit'. *Oikos* **1999**, *87*, 615–621. [CrossRef]
40. Lu, J.-N. The Effects of Arbuscular Mycorrhizal Fungi on the Growth of Three Grassland Plants. Master's Thesis, Hebei Agricultural University, Baoding, China, 2023.
41. Ren, Z.; Xia, T.Y.; Chen, L.J.; Han, L.; Chen, Z.B.; Bai, H.L. Effect of different AMF on physiological related indexes of corn. *Southwest China J. Agric. Sci.* **2015**, *28*, 563–568.
42. Ramírez-Flores, M.R.; Perez-Limon, S.; Li, M.; Barrales-Gamez, B.; Albinsky, D.; Paszkowski, U.; Olalde-Portugal, V.; Sawers, R.J. The genetic architecture of host response reveals the importance of arbuscular mycorrhizae to maize cultivation. *eLIFE* **2020**, *9*, e61701. [CrossRef]
43. Liu, R.C.; Yang, L.; Zou, Y.N.; Wu, Q.S. Root-associated endophytic fungi modulate endogenous auxin and cytokinin levels to improve plant biomass and root morphology of trifoliate orange. *Hortic. Plant J.* **2023**, *9*, 463–472. [CrossRef]
44. Mathur, S.; Sharma, M.P.; Jajoo, A. Improved photosynthetic efficacy of maize (*Zea mays* L.) plants with arbuscular mycorrhizal fungi (AMF) under high temperature stress. *J. Photochem. Photobiol. B Biol.* **2018**, *180*, 149–154. [CrossRef]
45. Chen, F.; Yan, S.; Wang, H.L.; Zhang, K.; Zhao, F.N.; Huang, X.Y. Study on gas exchange parameters and water use efficiency of spring wheat leaves under different levels of water stress. *Arid. Zone Res.* **2021**, *38*, 821–832.
46. Singh, M.; Sharma, J.G.; Giri, B. Augmentative role of arbuscular mycorrhizal fungi, *Piriformospora indica*, and plant growth-promoting bacteria in mitigating salinity stress in maize (*Zea mays* L.). *J. Plant Growth Regul.* **2024**, *43*, 1195–1215. [CrossRef]
47. Zhu, X.C.; Song, F.B.; Liu, S.Q.; Liu, T.D. Effects of arbuscular mycorrhizal fungus on photosynthesis and water status of maize under high temperature stress. *Plant Soil* **2011**, *346*, 189–199. [CrossRef]
48. Augé, R.M.; Toler, H.D.; Saxton, A.M. Arbuscular mycorrhizal symbiosis alters stomatal conductance of host plants more under drought than under amply watered conditions: A meta-analysis. *Mycorrhiza* **2015**, *25*, 13–24. [CrossRef] [PubMed]

49. Boldt, K.; Pörs, Y.; Haupt, B.; Bitterlich, M.; Kühn, C.; Grimm, B.; Franken, P. Photochemical processes, carbon assimilation and RNA accumulation of sucrose transporter genes in tomato arbuscular mycorrhiza. *J. Plant Physiol.* **2011**, *168*, 1256–1263. [CrossRef]
50. Liu, R.C.; Gao, W.Q.; Srivastava, A.K.; Zou, Y.N.; Kuča, K.; Hashem, A.; Abd_Allah, E.F.; Wu, Q.S. Differential effects of exogenous glomalin-related soil proteins on plant growth of trifoliate orange through regulating auxin changes. *Front. Plant Sci.* **2021**, *12*, 745402. [CrossRef]
51. Holátko, J.; Brtnický, M.; Kučerík, J.; Kotianová, M.; Elbl, J.; Kintl, A.; Kynický, J.; Benada, O.; Datta, R.; Jansa, J. Glomalin–truths, myths, and the future of this elusive soil glycoprotein. *Soil Biol. Biochem.* **2021**, *153*, 108116. [CrossRef]
52. Wu, Q.S.; Srivastava, A.K.; Li, Y. Effects of mycorrhizal symbiosis on growth behavior and carbohydrate metabolism of trifoliate orange under different substrate P levels. *J. Plant Growth Regul.* **2015**, *34*, 499–508. [CrossRef]
53. Salmeron-Santiago, I.A.; Martínez-Trujillo, M.; Valdez-Alarcón, J.J.; Pedraza-Santos, M.E.; Santoyo, G.; Pozo, M.J.; Chávez-Bárcenas, A.T. An updated review on the modulation of carbon partitioning and allocation in arbuscular mycorrhizal plants *Microorganisms* **2021**, *10*, 75. [CrossRef]
54. Göbel, M.; Fichtner, F. Functions of sucrose and trehalose 6-phosphate in controlling plant development. *J. Plant Physiol.* **2023**, *291*, 154140. [CrossRef]
55. Du, Y.L.; Zhao, Q.; Chen, L.R.; Yao, X.D.; Zhang, W.; Zhang, B.; Xie, F. Effect of drought stress on sugar metabolism in leaves and roots of soybean seedlings. *Plant Physiol. Biochem.* **2020**, *146*, 1–12. [CrossRef]
56. Eom, J.S.; Chen, L.Q.; Sosso, D.; Julius, B.T.; Lin, I.W.; Qu, X.Q.; Braun, D.M.; Frommer, W.B. SWEETs, transporters for intracellular and intercellular sugar translocation. *Curr. Opin. Plant Biol.* **2015**, *25*, 53–62. [CrossRef]
57. Feng, L.; Frommer, W.B. Structure and function of SemiSWEET and SWEET sugar transporters. *Trends Biochem. Sci.* **2015**, *40*, 480–486. [CrossRef] [PubMed]
58. Sugiyama, A.; Saida, Y.; Yoshimizu, M.; Takanashi, K.; Sosso, D.; Frommer, W.B.; Yazaki, K. Molecular characterization of *LjSWEET3*, a sugar transporter in nodules of *Lotus japonicus*. *Plant Cell Physiol.* **2017**, *58*, 298–306. [PubMed]
59. Yang, J.; Luo, D.; Yang, B.; Frommer, W.B.; Eom, J.S. *SWEET 11* and *15* as key players in seed filling in rice. *New Phytol.* **2018**, *218*, 604–615. [CrossRef]
60. Liu, X.; Zhang, Y.; Yang, C.; Tian, Z.; Li, J. *AtSWEET4*, a hexose facilitator, mediates sugar transport to axial sinks and affects plant development. *Sci. Rep.* **2016**, *6*, 24563. [CrossRef]
61. Kryukov, A.A.; Gorbunova, A.O.; Kudriashova, T.R.; Yakhin, O.I.; Lubyanov, A.A.; Malikov, U.M.; Shishova, M.F.; Kozhemyakov, A.P.; Yurkov, A.P. Sugar transporters of the *SWEET* family and their role in arbuscular mycorrhiza. *Vavilov J. Genet. Breed.* **2021**, *25*, 754. [CrossRef] [PubMed]
62. Jiang, Y.; Wang, W.; Xie, Q.; Liu, N.A.; Liu, L.; Wang, D.; Zhang, X.; Yang, C.; Chen, X.; Tang, D. Plants transfer lipids to sustain colonization by mutualistic mycorrhizal and parasitic fungi. *Science* **2017**, *356*, 1172–1175. [CrossRef]
63. Li, Q.S.; Srivastava, A.K.; Zou, Y.N.; Wu, Q.S. Field inoculation responses of arbuscular mycorrhizal fungi versus endophytic fungi on sugar metabolism associated changes in fruit quality of lane late navel orange. *Sci. Hortic.* **2023**, *308*, 111587. [CrossRef]
64. Rich, M.K.; Vigneron, N.; Libourel, C.; Keller, J.; Xue, L.; Hajheidari, M.; Radhakrishnan, G.V.; Le Ru, A.; Diop, S.I.; Potente, G.; et al. Lipid exchanges drove the evolution of mutualism during plant terrestrialization. *Science* **2021**, *372*, 864–868. [CrossRef]

agriculture

MDPI

*Article*

# Selection and Characterisation of Elite *Mesorhizobium* spp. Strains That Mitigate the Impact of Drought Stress on Chickpea

**María Camacho [1,†], Francesca Vaccaro [2,†], Pilar Brun [1], Francisco Javier Ollero [3], Francisco Pérez-Montaño [3], Miriam Negussu [2], Federico Martinelli [2,*], Alessio Mengoni [2], Dulce Nombre Rodriguez-Navarro [1] and Camilla Fagorzi [2,*]**

1 IFAPA Centro Las Torres Tomejil, 41200 Sevilla, Spain; mariag.camachomartinez@juntadeandalucia.es (M.C.); dulcenombre.rodriguez@juntadeandalucia.es (D.N.R.-N.)
2 Department of Biology, University of Florence, 50019 Sesto Fiorentino, Italy; alessio.mengoni@unifi.it (A.M.)
3 Department of Microbiology, Faculty of Biology, Universidad de Sevilla, 41004 Seville, Spain; fjom@us.es (F.J.O.); fperezm@us.es (F.P.-M.)
* Correspondence: federico.martinelli@unifi.it (F.M.); camilla.fagorzi@unifi.it (C.F.)
† These authors contributed equally to this work.

**Abstract:** The chickpea (*Cicer arietinum* L.) is a key legume crop in Mediterranean agriculture, valued for its nutritional profile and adaptability. However, its productivity is severely impacted by drought stress. To identify microbial solutions that enhance drought resilience, we isolated seven *Mesorhizobium* strains from chickpea nodules collected in southern Spain and evaluated their cultivar-specific symbiotic performance. Two commercial cultivars (Pedrosillano and Blanco Lechoso) and twenty chickpea germplasms were tested under growth chamber and greenhouse conditions, both with and without drought stress. Initial screening in a sterile substrate using nodulation assays, shoot/root dry weight measurements, and acetylene reduction assays identified three elite strains (ISC11, ISC15, and ISC25) with superior symbiotic performance and nitrogenase activity. Greenhouse trials under reduced irrigation demonstrated that several strain–cultivar combinations significantly mitigated drought effects on plant biomass, with specific interactions (e.g., ISC25 with RR-98 or BT6-19) preserving over 70% of shoot biomass relative to controls. Whole-genome sequencing of the elite strains revealed diverse taxonomic affiliations—ISC11 as *Mesorhizobium ciceri*, ISC15 as *Mesorhizobium mediterraneum*, and ISC25 likely representing a novel species. Genome mining identified plant growth-promoting traits including ACC deaminase genes (in ISC11 and ISC25) and genes coding for auxin biosynthesis-related enzymes. Our findings highlight the potential of targeted rhizobial inoculants tailored to chickpea cultivars to improve crop performance under water-limiting conditions.

**Keywords:** *Mesorhizobium* spp.; chickpea; drought stress; plant-microbe interaction; nitrogen fixation; bioinoculants; genotype-by-genotype interaction

## 1. Introduction

The chickpea (*Cicer arietinum* L.) is an essential legume crop widely cultivated for its high nutritional value and adaptability to various environmental conditions. This legume is rich in proteins, fibres, and minerals, making it a valuable dietary component in many regions, especially in the Mediterranean basin [1]. Recent studies indicate that doubling plant-based foods while cutting back on meat and sugar supports both human and planetary health in a climate-challenged world. [2]. Increasing legume consumption also provides significant environmental benefits, as their cultivation enhances soil fertility, boosts nitrogen levels and organic matter content, and supports key ecosystem services—such as improved resilience to biotic stress, reduced reliance on pesticides and fertilizers, and greater support for pollinators. Therefore, it is essential to develop new knowledge and innovative biotechnological tools to fully utilize and enhance legume cultivation, addressing the growing challenges of drought and salinity exacerbated by climate change [3–5]. In this sense, the reduction in water availability poses a significant threat to legumes, as evidenced by the 60% decline in bean production under rainfed conditions globally and the 80% reduction in grain yield in certain arid regions [6]. Drought stress not only limits water uptake but also disrupts both short-term and long-term adaptation mechanisms in plant species facing climate change. Drought-resistant cultivars, even within the same species, exhibit an advantageous strategy by regulating carbohydrate allocation towards seed filling, helping to mitigate the impact of drought during the pod filling stage [5,6]. Moreover, the chickpea, like other legume crops, is key in low-input (sustainable) agriculture, since it forms a nitrogen-fixing symbiotic association with bacteria called rhizobia. This symbiosis establishes in specific root structures, called nodules, which are colonized by bacteria. Bacteria colonizing nodules penetrate into plant cells and differentiate into a form called bacteroid, where nitrogen fixation (the formation of ammonia from atmospheric dinitrogen) takes place [7,8]. Under this stressed condition, the inoculation with selected rhizobial strains can alleviate plant stress, hence playing a relevant role in plant tolerance. Data from the literature has shown that the specificity between rhizobia strains and plant species has been largely affected by the relative occurrence and distribution of the potential symbionts in the soils of different regions [9]. A vast characterization of chickpea rhizobia has been conducted in the last three decades in Europe, Africa, Asia, and Oceania [10]. Chickpea has traditionally been considered as a restrictive host for nodulation [7]. Classically, four rhizobia species belonging to the genus *Mesorhizobium* have been identified as chickpea symbionts, each with distinct characteristics and ecological adaptations [11]. *M. ciceri* was the first species described, isolated from chickpea nodules in Spain, and is known for its effectiveness in the neutral to slightly alkaline soils typical of Mediterranean regions [11]. Following this, *M. mediterraneum* was described, also from the Mediterranean region, distinguished by its ability to grow in slightly acidic conditions and adapt to drier soils with higher temperatures [12]. Interestingly, *M. ciceri* and *M. mediterraneum* have been found in chickpea nodules across many countries such as Spain, Portugal, Morocco, Tunisia, and India [7] but not in China where two other *Mesorhizobium* species were isolated from chickpea root nodules from saline–alkaline soils and were identified as *Mesorhizobium. muleiense* and *Mesorhizobium wenxiniae* [13,14]. Moreover, other species can nodulate chickpea such as *Mesorhizobium amorphae*, *Mesorhizobium loti*, *Mesorhizobium tianshanense*, *Mesorhizobium oportunistum*, *Mesorhizobium abyssinicae*, and *Mesorhizobium shonense* [7,15–20]. Additionally, rhizobia from other genera, such as *Ensifer* nodulating chickpea, have also been reported [21,22]. These works highlighted the high diversity in rhizobia strains of chickpea nodules and its possible agronomic effect on different agro-ecosystems and cultivation types and environments. However, selecting the correct rhizobial strain to be used as an inoculant is not an easy task. *Mesorhizobium* strains could contribute to plant growth promotion through multiple mechanisms. These bacteria may produce plant growth-promoting substances like auxins, cytokinins, and gibberellins, which enhance root and shoot development, nutrient uptake, and overall plant vigour [21–23]. In this sense, the enhanced root system improves the plant's ability to acquire water and nutrients from the soil under starvation, thereby increasing its resilience to abiotic stresses. On the other hand, rhizobia strains expressing the 1-aminocyclopropane-1-carboxylate (ACC) deaminase enzyme have been reported to display an augmented symbiotic performance as a consequence of lowering the plant ethylene levels that inhibit the nodulation process during different stresses, such as drought conditions [24]. The selection of efficient *Mesorhizobium* strains that can effectively nodulate chickpea and sustain productivity under drought is, therefore, of great agronomic interest. However, it is important to note that the success of chickpea inoculation with *Mesorhizobium* strains is influenced by strain specificity and host compatibility. Thus, different *Mesorhizobium* strains exhibit varying degrees of compatibility with chickpea cultivars, which means that some strains are more effective in promoting nitrogen fixation and plant growth in certain chickpea varieties compared with others [8]. This emphasizes the significance of strain selection based on the specific chickpea cultivar and local soil conditions. Here, we aimed to identify and evaluate the plant cultivar-specific performance of a collection of *Mesorhizobium* strains in promoting chickpea growth under drought-stressed conditions. A combination of in vitro and greenhouse symbiotic tests with different chickpea cultivars and complete genome sequencing of three top-performing *Mesorhizobium* strains were performed to provide evidence for the potential use of these strains as elite inoculants for selected chickpea cultivars under water-limiting conditions.

## 2. Materials and Methods

### 2.1. Bacterial Strains, Culture Conditions, and Chickpea Cultivars

*Mesorhizobium* ISC strains were isolated from *Cicer arietinum* root nodules from different geographic areas of the south of Spain (Table 1) following the methodology described by Somasegaran and Hobben [25]. Briefly, surface-sterilized nodules were crushed on mannitol agar (YMA) medium. Then, the isolates were purified by repeated streaking on the same medium. These strains were grown at 28 °C on tryptone yeast (TY) medium [26] or modified yeast extract mannitol (YMB) medium [27]. Pure cultures were kept at −80 °C in YMB plus glycerol (1:1 *vol:vol*).

**Table 1.** *Mesorhizobium* spp. strains used in this work. Strain codes, cultivar of isolation, and geographical location in Spain of nodulated chickpea plants are reported.

| Strain Denomination | Cultivar of Isolation | Origin (Location and Province) |
|---|---|---|
| *ISC6* | Pedrosillano | Carmona (Sevilla) |
| *ISC11* | Non determined | Holguera (Cáceres) |
| *ISC13* | Non determined | Cañaveral (Cáceres) |
| *ISC15* | Pedrosillano | La Roca (Badajoz) |
| *ISC18* | Garbanzo de Valencia del Ventoso | Fuente de Cantos (Badajoz) |
| *ISC24* | Pedrosillano | Carmona (Sevilla) |
| *ISC25* | Blanco lechoso | Huelva (Huelva) |

The majority of *Cicer arietinum* germplasms used in this work are recombinant inbred lines described previously [1], with three of them being registered cultivars. Two reference cultivars, Pedrosillano and Blanco Lechoso, were included for selecting the elite *Mesorhizobium* strains (Table 2).

### 2.2. Symbiotic Assays in Growth Chambers

Two experimental approaches were carried out in symbiotic assays in growth chambers.

Firstly, for the evaluation of the symbiotic phenotypes, the seven *Mesorhizobium* ISC strains were grown in YMB medium until the bacterial concentration reached about $10^9$ cells $mL^{-1}$. Bacterial concentration was checked by viable count of the cultures. Seeds of chickpea cv Pedrosillano and Blanco Lechoso were surface-sterilized for 5′ in a 3% $Ca(ClO)_2$ solution and pre-germinated and placed in sterilized jars (Leonard jars) containing vermiculite and perlite substrates (3:1) and Farhäeus N-free solution, as previously described [28,29]. Each pre-germinated and sterilized seed was inoculated with 1 mL of bacterial culture. Growth conditions were 16 h at 26 °C in the light and 8 h at 18 °C in the dark, with 70% humidity. Nodulation test parameters (shoot and root dry weight, nodule number, and dry weight of nodules) were evaluated after 4 weeks. Nitrogenase activity in the nodules was estimated by acetylene reduction assays (ARAs) as described by Buendía-Clavería et al. [30]. Nodulation test experiments were replicated four times.

**Table 2.** Chickpea germplasms used in this work and their origin.

| Cultivar | Type and Origin |
|---|---|
| *Pedrosillano* | Commercial cultivar, Salamanca (Spain) |
| *Blanco Lechoso* | Commercial cultivar, Andalucía (Spain) |
| *5-RIL33* | Recombinant inbred lines |
| *5-RIL92* | Recombinant inbred lines |
| *RR-33 Valeka* | Recombinant inbred lines, cultivar, Andalucía (Spain) |
| *RR-51* | Recombinant inbred lines |
| *RR-98 Kasin* | Recombinant inbred lines, cultivar, Andalucía (Spain) |
| *BT3-13* | Recombinant inbred lines |
| *BT3-23* | Recombinant inbred lines |
| *BT5-7* | Recombinant inbred lines |
| *BT6-17 Kaveri* | Recombinant inbred lines, cultivar, Andalucía (Spain) |
| *BT6-19* | Recombinant inbred lines |
| *22-2M2/1-4-3 (g_8)* | Mutagenic Pascià |
| *4 ENEA/2-5-1 (g_11)* | Population Italy |
| *6-1M2/4-3-5(g_14)* | Mutagenic Pascià |
| *GSC-21-2 (g_20)* | Population Italy |
| *GSC-30-1 (g_21)* | Population Italy |
| *GSC-5-2 (g_29)* | Population Italy |
| *GSC-35-3 (g_31)* | Population Italy |
| *GSC-16-3 (g_33)* | Population Italy |
| *FLIP03-112C (g_41)* | Recombinant inbred lines (X00TH51/FLIP98-52CxFLIP98-47C) |
| *FLIP10-125C (g_45)* | Recombinant inbred lines (X04TH-85/X03TH-153XS01114) |

Secondly, chickpea seeds (Germplasms g_8, 11, 14, 20, 21, 29, 31, 33, 41, and 45, listed in Table 2) were surface-sterilized in 0.25% $HgCl_2$ solution for 2 min and washed five times with sterile water. After the sterilization procedure, seeds were kept in a Petri dish in the dark at 26 °C until germination, then seedlings were placed in sterile polypropylene jars containing vermiculite/perlite (1:1) and nitrogen-free Farhäeus solution, and grown at 26 °C with a 16 h photoperiod. One-week-old plantlets were inoculated with $10^5$ $mL^{-1}$ appropriate *Mesorhizobium* ISC strain cells washed twice in NaCl 0.9%. Ten plants of the same cultivar were inoculated per strain and then grown for 5 weeks in the same conditions. After 5 weeks, the plants were harvested, and the shoot and root dry weights were measured, as reported also in Checcucci et al., 2018 [31]. To evaluate the influence of strains and plant genotypes, as well as the interaction of both factors, aerial part length and dry weight, a two-way permanova was run in R (v 4.4.1) using the 'adonis2' function of the Vegan package [32].

### 2.3. Symbiotic Assays in Greenhouse Under Drought Conditions

Plastic pots of 250 mL were filled up with a mixture of perlite/vermiculite (1:2). Chickpea seeds were surface-disinfected with $Ca(ClO)_2$ for 5 min, rinsed with sterilized water at least five times, and germinated in agar/water (1%) at 28 °C. The test plant involved 10 chickpea genotypes (cv. 5-RIL33,5-RIL92, RR-33, RR-51,RR-98,BT3-13, BT3-23, BT5-7, BT6-17, and BT6-19), three mesorhizobia strains (*Mesorhizobium* spp. ISC11, ISC15 and ISC25), and two water regimes: full irrigation (control), with 200 mL water twice a week, and 30% of the control. For watering, 50% Rigaud and Puppo (1979) nutrient solution without nitrogen (N-free) was employed [33]. Pots were set up under greenhouse conditions, sowing with two pre-germinated seeds per plot. Four replicates were arranged per genotype × mesorhizobia × water regime combination. At the time of sowing, seeds were inoculated with 1 mL of a YMB stationary culture with about $10^9$ cells $mL^{-1}$ of the corresponding strain. After 4 weeks, plants were harvested and the following biometric measurements were performed: shoot and root dry weight, nodule number, and nodule fresh weight.

### *2.4. Genome Sequences and Annotation*

Prior isolating total genomic DNA, the strains were grown to stationary phase at 30 °C in TY medium. DNA isolation was performed with a DNeasy® PowerLyzer® PowerSoil® Kit (Qiagen, Hessen, Germany). DNA quality control and long read sequencing were performed by the service company Macrogen Europe (Macrogen Europe, Amsterdam, The Netherlands) using a Pacific Biosciences Sequel II instrument (Pacific Biosciences, Menlo Park, CA, USA). Obtained reads were analyzed using HGAP v. Microbial Assembly 8 within the SMRT Link software ver. 8.0.0.80529 (Pacific Biosciences, Menlo Park, CA, USA) producing an *oriC*-oriented assembly by running the "microbial multiplexing" pipeline with default options. Annotation was performed by the NCBI Prokaryotic Genome Annotation Pipeline (PGAP) [34].

### *2.5. Genome Characterization and Mining*

The taxonomic identification of strains from the genome sequence was performed using the Type (Strain) Genome Server (TYGS) [35]. Digital DNA–DNA hybridization (dDDH) values were computed with the Genome BLAST Distance Phylogeny approach. Formula *d4* was used since it is independent of genome length and is thus robust against the use of incomplete draft genomes. The mapping of Kyoto Encyclopedia of Genes and Genomes (KEGG) pathways was performed using the KEGG Automatic Annotation Server (KAAS) on the KEGG web server [36]. Secondary metabolite gene clusters have been identified by AntiSMASH [37]. The Plant Growth-Promoting Traits Prediction tool (PGPT-Pred) from PLaBAse (https://plabase.cs.uni-tuebingen.de/, accessed on 29 July 2025 [38]) was used to allow genomic protein annotation (blastp + hmmer approach) mapping against the PGPT ontology (approx. 6900 bacterial plant growth-promoting traits). Pangenome analysis with Roary v3.13.0 was used to define core and dispensable/unique gene fractions against close relatives of each strain. For this purpose, sequences of close relative strains (*Mesorhizobium*-type strains belonging to the same groups of the three strains of the current work, on the basis of the phylogenetic tree obtained from TYGS analysis) were downloaded from the NCBI Genome Database. Genomes were reannotated with Prokka to ensure consistent annotation [39]. All genomes were downloaded on 8 December 2023. The three pangenomes of the *Mesorhizobium* strains were calculated using Roary 3.13 with a percentage identify threshold of 95% [40].

### *2.6. Identification and Phylogenetic Analysis of Common NodC and NifH Proteins*

The sequences of NodC and NifH proteins were searched for in the Prokka Genome Annotation files of the three strains. Each protein sequence was provided as input to BLASTP against the Antimicrobial Resistance (AMR) protein database limited to *Mesorhizobium* sp (taxid:1871066). The amino acid sequences for each hit (regardless of e-value) belonging to *Mesorhizobium* spp.-type strains were collected. Alignments and phylogenies were built using MEGA11 [41]. The evolutionary history was inferred by using the maximum likelihood method and Jones–Taylor–Thornton (JTT) matrix-based model [42]. The tree is drawn to scale, with branch lengths measured in the number of substitutions per site.

## 3. Results

### 3.1. Phenotypic Analysis of Commercial Chickpea-Nodulating Strains Identifies ISC11, ISC15, and ISC25 as Elite Strains

Ref. [29] was used to determine which of the seven Mesorhizobium ISC strains were elite rhizobia for nodulating with this legume. We studied their symbiotic behaviours with the two main commercial *C. arietinum* varieties sown in this Mediterranean country, namely Pedrosillano and Blanco Lechoso. Three different parameters were analyzed for each symbiotic partner: shoot dry weight, number of nodules, and nodule dry weight (Table 3). The best symbiotic performances in the Pedrosillano chickpea germplasm were displayed by strain ISC25, followed by strains ISC11, ISC15, and ISC6. Conversely, when seedlings of chickpea cv. Blanco Lechoso were inoculated with the same *Mesorhizobium* ISC strains, ISC11 showed the best values in the symbiotic parameters analyzed, especially in shoot dry weight, followed by ISC15. Additionally, the *Mesorhizobium* strain ISC25 also displayed an improved symbiotic performance, although the results were not significantly different to those obtained by strains ISC6, ISC18, and ISC24. Thus, three of the *Mesorhizobium* strains, ISC11, ISC15, and ISC25, were selected as elite rhizobia and inspected to determine their nitrogenase activity (estimated by ARA) in nodulation assays in Leonard jars with both chickpea commercial germplasms (Figure 1). According to our results, the best acetylene reduction abilities were found in nodules colonized by ISC11 or ISC15 in the case of Pedrosillano germplasm, and ISC11 for Blanco lechoso (Figure 1).

**Table 3.** Symbiotic responses of chickpea commercial cultivars. The symbiotic quality of inoculation with ISC strains with Pedrosillano and Blanco lechoso chickpea varieties are reported in terms of number and dry weight of nodules and plant and shoot dry weight. Values are means of 4 independent replicates. In the same column, for each cultivar, data followed by the same letter are not significantly different (one-way ANOVA, $p < 0.05$).

| Strains | Number of Nodules | Nodule Dry Weight (g) | Shoot Dry Weight (g) |
|---|---|---|---|
| *Pedrosillano* | | | |
| *ISC6* | 91.0 ± 11.0 ab | 414.6 ± bc | 3.1 ± 0.3 ab |
| *ISC11* | 119.5 ± 4.5 a | 553.8 ± 70.0 ab | 4.1 ± 0.8 ab |
| *ISC13* | 85.0 ± 10.0 b | 122.8 ± 14.2 d | 0.6 ± 0.1 c |
| *ISC15* | 73.0 ± 3.4 bc | 416.5 ± 68.0 bc | 3.1 ± 0.5 ab |
| *ISC18* | 73.8 ± 8.0 bc | 274.0 ± 46.6 cd | 2.3 ± 0.6 bc |
| *ISC24* | 72.5 ± 14.9 bc | 251.6 ± 12.4 cd | 2.6 ± 0.7 abc |
| *ISC25* | 122 ± 10.1 a | 663.3 ± 140.8 a | 4.6 ± 1.2 a |
| *Non-inoculated* | 0 | 0 | 0.6 ± 0.1 c |
| *Blanco lechoso* | | | |
| *ISC6* | 57.3 ± 39 ab | 232 ± 11 ab | 2.7 ± 0.6 bcd |
| *ISC11* | 38.0 ± 24 b | 132.0 ± 6 b | 5.0 ± 1.3 a |
| *ISC13* | 101.0 ± 17 a | 201 ± 111 ab | 2.1 ± 0.6 cd |
| *ISC15* | 41 ± 10 b | 299 ± 77 a | 3.5 ± 0.4 b |
| *ISC18* | 21.0 ± 4 b | 136 ± 133 b | 2.6 ± 0.5 bcd |
| *ISC24* | 48.0 ± 15 b | 133 ± 17 b | 2.2 ± 0.1 bcd |
| *ISC25* | 19.0 ± 7 b | 190 ± 42 ab | 3.4 ± 0.4 bc |
| *Non-inoculated* | 0 | 0 | 1.5 ± 0.2 d |

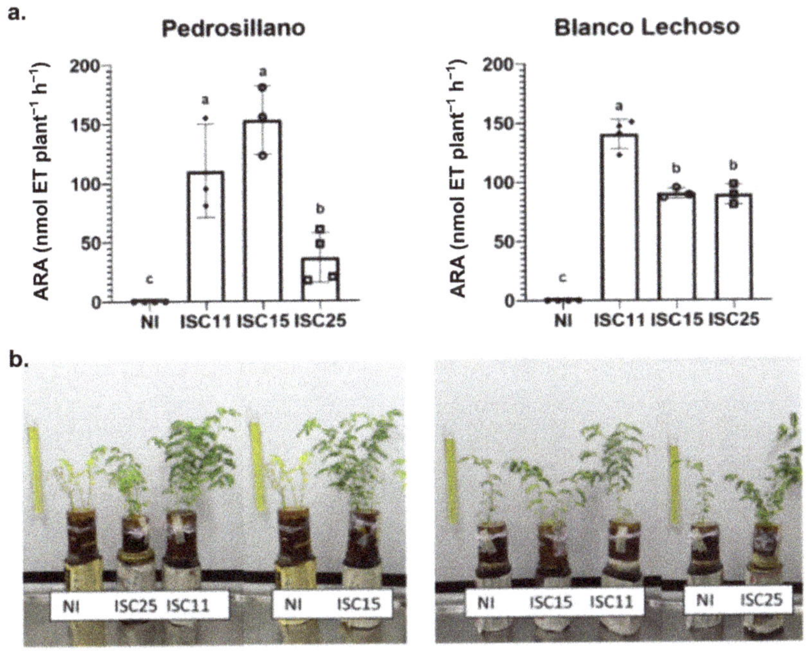


**Figure 1.** Estimation of nitrogenase performance of *Mesorhizobium* elite strains. Results of the acetylene reduction assays (ARAs) on chickpea cv. Pedrosillano (**left**) and Blanco Lechoso (**right**) inoculated with ISC elite strains. (**a**) ARA per Leonard jar. Error bars show standard deviations of ISC11, ISC15, and ISC25 elite strains. Bars marked by the same letter are not significantly different (one-way ANOVA, $p < 0.05$). (**b**) Representative plants of each condition. NI, non-incolulated.

### *3.2. Genome Sequencing and Taxonomic Assignment of* Mesorhizobium *Elite Strains*

The genome sizes for the three strains ranged from ~6.7 Mbp to ~7.3 Mbp, composed of a large chromosome and one or two plasmids (Table 4).

**Table 4.** Genome sequencing statistics and annotation. Genome size, number of contigs, and the number of coding sequences (CDS), ribosomal RNA (rRNA), transfer RNA (tRNA), and transfer-messenger RNA (tmRNA) genes are reported.

| | *ISC11* | *ISC15* | *ISC25* |
|---|---|---|---|
| *Genome size (bp)* | 6,719,922 | 6,908,843 | 7,259,116 |
| *n. of contigs/replicons* | 2 | 2 | 3 |
| *CDS* | 6612 | 6599 | 6961 |
| *rRNA* | 3 | 6 | 6 |
| *tRNA* | 51 | 54 | 52 |
| *tmRNA* | 1 | 1 | 1 |

For each genome, in silico DNA–DNA hybridization (DDH) and genome-based phylogeny were computed to allow a robust taxonomic classification of the strains [33]. The results of the taxonomic assignment based on dDDH values are reported in Supplementary Dataset S1 and the genome-based phylogenetic tree is shown in Figure 2. The strains ISC11 and ISC15 belonged to the already-described species, *M. ciceri* and *M. mediterraneum*, respectively, which were already known as symbionts of chickpea. On the contrary, ISC25 showed only low dDDH values with known species (~30–40%), thereby supporting the hypothesis that it could constitute a novel species of chickpea-nodulating rhizobia.

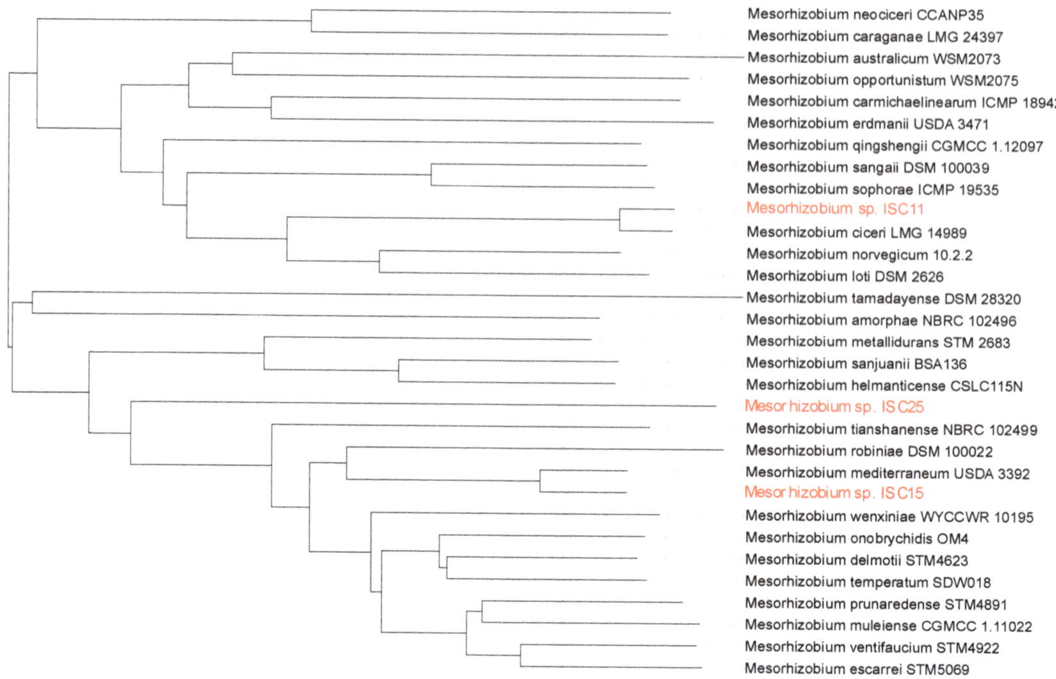


**Figure 2.** Genome-based taxonomy of elite ISC strains. Genome-based evolutionary analysis by maximum likelihood method of *Mesorhizobium* ISC11, ISC15, and ISC25 strains. Tree inferred with FastME 2.1.6.1 from Genome BLAST Distance Phylogeny (GBDP) distances calculated from genome sequences [43]. The branch lengths are scaled in terms of GBDP distance formula d5. The numbers above branches are GBDP pseudo-bootstrap support values > 60% from 100 replications, with an average branch support of 96.5%.

To better evaluate the similarities between strains, we calculated the number of core and dispensable genes between the strains ISC11, ISC15, and ISC25 and their close relatives. The Roary computational pipeline was run for three groups of strains: group 1 (*Mesorhizobium* sp. ISC 11 and *M. ciceri*), group 2 (*Mesorhizobium* sp. ISC 15 and *M. mediterraneum*), and group 3 (*Mesorhizobium* sp. ISC 25, *M. thianshianense, M. mediterraneum, M. onobrychidis, M. delmotii, M. temperatum, M. muleiense, M. ventifaucium,* and *M. escarrei*). Strains ISC11 and ISC15 share, respectively, 27% and 33% of their genes with their close relatives *M. ciceri* and *M. mediterraneum*. ISC25 share 22% of genes with the strains belonging to its group (Supplementary Dataset S2). Based on these data we could assign ISC11 to *M. cicero* and ISC15 to *M. mediterraneum*, while ISC25 could deserve future taxonomic evaluation as a possible novel species.

Given the relevance to symbiotic nitrogen fixation and the genome diversity of strains, we checked for the presence of the common set of genes required for Nod-factor-dependent symbiotic interaction (*nod* genes) and nitrogen fixation (*nif* genes) and we then inspected the phylogeny of NodC and NifH to establish a possible common origin (Figure S1a,b). The results indicated that ISC11 and ISC15 share identical NifH proteins, clustering together with the *M. mediterraneum* ortholog. For NodC, ISC15 and ISC25 share the same sequence, which is very close to the ISC11 and *M. ciceri* ortholog. These results suggest that, though being genomically different species, these strains experienced some level of horizontal gene transfer of symbiotic elements.

To further inspect the presence of other genes potentially involved in plant growth promotion, we reannotate the genomes of ISC11, ISC15, and ISC25 using the PLaBAse ontology tool for plant growth-promoting traits. The results (Table 5, Supplementary Dataset S3) indicated for all strains the putative presence of several traits, other than symbiotic nitrogen fixation, which can help plant growth. To further check the presence of specific genes related to plant growth we analyzed the presence of the *acdS* gene, encoding for the ACC deaminase (1-aminocyclopropane-1-carboxylate deaminase) enzyme that is involved in reducing the ethylene levels in the plant, and the presence of pathways for the production of the phytohormones auxin indole-3-acetic acid (IAA) and cytokinin zeatin (Supplementary Material File, Figure S2). The strains ISC11 and ISC25 harboured the *acdS* gene, while no *acdS* gene was found in the genome of ISC15. Notably, ISC25 harboured two *acdS* genes, located in different genomic positions (Supplementary Material File, Table S1). Concerning potential IAA production, ISC11 harboured enzyme-converting indel-3-acetaldeyde to indole-3-acetate, though apparently no reaction for the production of such substrates seems to be present. ISC15 and ISC25 also harboured some enzymes for reactions involving IAA, but similarly to ISC11 no direct connection with tryptophan metabolism is present. The pathway for zeatin production was absent from all strains.

**Table 5.** Number of genes of *Mesorhizobium* ISC11, ISC15, and ISC25 strains related to the ontologies for plant growth-promoting traits. Data from blast + hmmer annotation against the PLaBAse database.

| | *ISC11* | *ISC15* | *ISC25* |
|---|---|---|---|
| *Biofertilization* | 12 | 11 | 11 |
| *Phytohormone/plant signal production* | 11 | 11 | 11 |
| *Plant immune response stimulation* | 1 | 1 | 1 |
| *Colonizing plant system* | 31 | 30 | 31 |
| *Competitive exclusion* | 18 | 20 | 19 |
| *Stress control/biocontrol* | 19 | 19 | 19 |
| *Bioremediation* | 7 | 7 | 7 |
| *Putative functions* | 0 | 0 | 0 |

Finally, in relation to secondary metabolite gene clusters, no clusters were identified in ISC11, while in ISC15 only a cluster for the biosynthesis of homoserine lactone (i.e., involved in quorum sensing) was found. On the contrary, for ISC25, four putative systems for quorum sensing were identified (homoserine lactone), one Type III polyketide synthase (PKS), plus other possible products (RiPP, terpene, cofactors) (Supplementary Material File, Table S2).

### 3.3. Variability in Symbiotic Quality on Chickpea Accessions

Given the high genomic diversity among the *Mesorhizobium* strains ISC11, ISC15, and ISC25, we further evaluated the extent of variability in symbiotic quality (as plant aerial part length and plant dry weight) under controlled conditions (plant growth chamber and inoculation in sterile substrate). In particular, we inspected the overall contribution of the strains, the host plant genotype (accession), and their interaction in symbiotic quality to determine the presence and the extent of a rhizobium genotype × plant genotype interaction. The results are reported in Figure 3 and Table S3. In general, the plant aerial part length showed a higher variability and ability to distinguish between accessions than plant dry weight, suggesting the hypothesis that differences could be due to the promotion of internodal separation than to the increase in biomass. The major effect of plant genotype on plant aerial part length was clear from the data shown in Table S3a. However, the *Mesorhizobium* strains also had a main effect on this parameter, as highlighted by the F statistics value. A statistically significant effect of the interaction between strain and accession was found for plant aerial length, indicating the presence of a genotype-by-genotype interaction. However, no effect on dry weight was observed (Table S3b). We cannot a priori exclude that testing in a pot and also using a larger set of accessions could result in evidence for genotype-by-genotype interaction also at the dry weight level and for the production and quality of seeds, as previously found in *S. meliloti* strains and varieties of the host plant alfalfa [44].

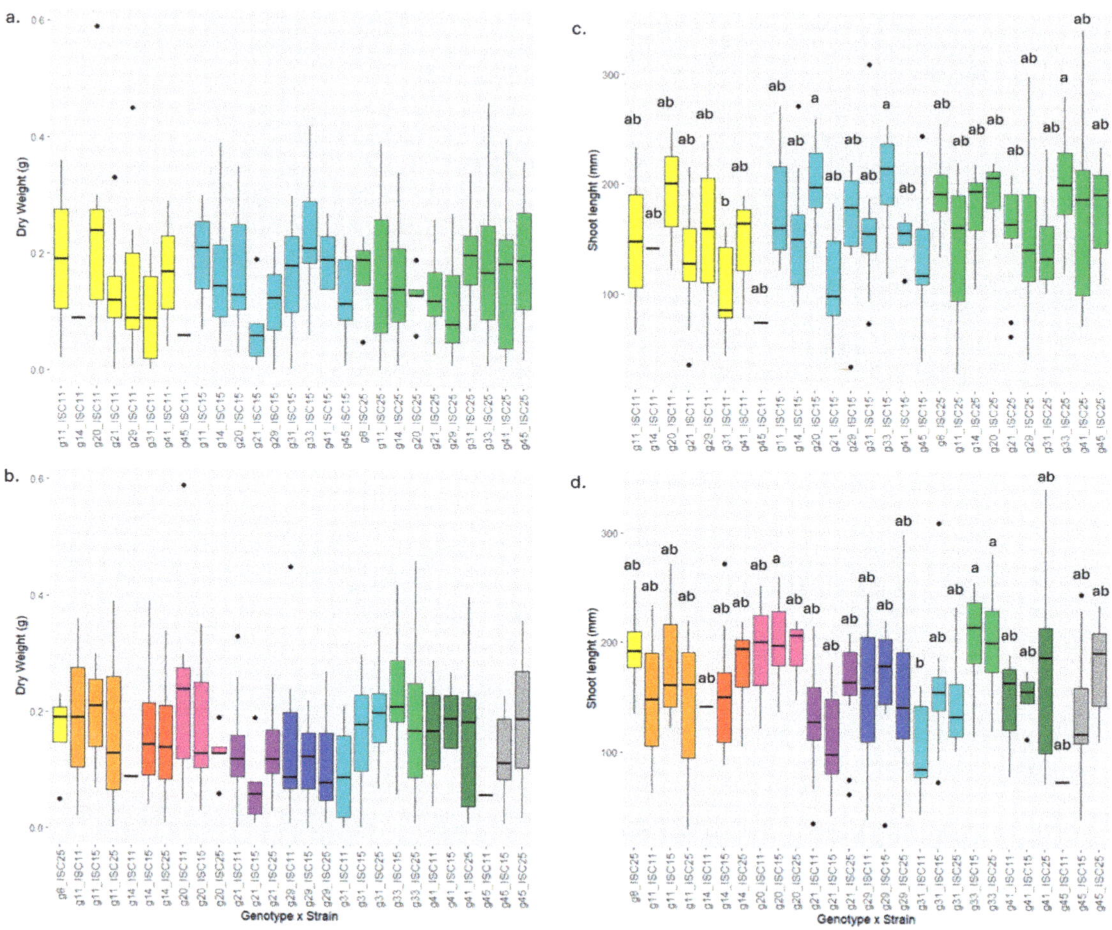


**Figure 3.** Variability in symbiotic quality among strains and chickpea accessions. Results from the in vitro test are reported. The figure shows the dry weight (**a**,**b**) and shoot length (**c**,**d**) measured for each plant inoculated with the different *Mesorhizobium* strains, harvested after 5 weeks of growth. Panels (**a**,**c**) evidence the effect of the inoculation with the three *Mesorhizobium* strains, while panels (**b**,**d**) show the effect of the ten plant genotypes. Multiple (pair-wise) comparisons were calculated using ANOVA and Tukey's test; the compact letter indicates different groups based on Tukey's test. No significant difference is obtained when measuring the dry weight of the plants—panels (**a**,**b**). Letters over the boxes in panels (**c**,**d**) show groups that are statistically different to each other.

*3.4. Mesorhizobium Elite Strains Differentially Support the Growth of Chickpea Cultivars under Water Scarcity*

The data regarding the interactions between plant genotype, rhizobia genotype, and irrigation level are summarized in Table 6 and displayed in detail in Figure 4 and Supplementary Dataset S4. The summarized data illustrate the percentage values for each parameter under water shortage conditions, contrasted with the control plants. According to the analysis of shoot dry weights, the following recommendations can be established to tackle drought conditions in the different germplasms, since the percentage of reduction was at least less than 40% with respect to the control (percentage values higher than 60). Thus, the *Mesorhizobium* ISC11 strain only mitigated the effect of water scarcity with the cultivars BT3-23 and BT6-19. In case of the elite strains ISC15 and ISC25, five and four chickpea germplasms for each strain (5-RIL33, RR-33, RR-98, BT13, and BT6-19; and RR-33, RR-51, RR-98, and BT6-19 Kaveri, respectively) were satisfactorily employed as buffers under this stressful condition. On the other hand, genotypes 5-RIL92, BT5-7, and BT6-17 did not perform well under water shortage conditions, with the shoot dry weight reduction ranging from 50% to 85% regardless of the presence of rhizobial strains. Conversely, the BT3-23 and BT6-19 cultivars, which were adversely affected by water scarcity in the absence of rhizobia, were significantly impacted by the presence of the three *Mesorhizobium* elite strains.

**Table 6.** Symbiotic parameters of chickpea genotypes under reduced watering. Percentage with respect to controls are reported. Root dry weight (RDW) and shoot dry weight (SDW) are reported. Treatments with a reduction higher than 60% (weight lower than 40%) in the SDW or RDW with respect to control conditions are highlighted in red. Values above 50 are indicated in red colour.

| ***Chickpea* Genotypes** | ***Mesorhizobium* Strains** | | | | | | | |
|---|---|---|---|---|---|---|---|---|
| | ISC11 | | ISC15 | | ISC25 | | Non-inoculated | |
| | SDW | RDW | SDW | RDW | SDW | RDW | SDW | RDW |
| *5-RIL33* | 44.6 | 86.3 | 68.0 | 76.6 | 49.5 | 72.7 | 38.2 | 71.8 |
| *5-RIL92* | 39.5 | 71.5 | 33.3 | 78.4 | 43.3 | 90.3 | 22.5 | 16.0 |
| *RR-33* | 44.1 | 66.0 | 69.7 | 98.3 | 64.0 | 88.0 | 23.7 | 11.0 |
| *RR-51* | 35.4 | 73.2 | 56.6 | 107.1 | 66.2 | 76.4 | 17.7 | 18.0 |
| *RR-98* | 30.4 | 57.4 | 62.5 | 86.6 | 70.4 | 78.1 | 18.9 | 22.3 |
| *BT-13* | 59.2 | 85.4 | 93.8 | 73.5 | 39.7 | 65.1 | 30.3 | 53.6 |
| *BT3-23* | 69.9 | 95.0 | 50.0 | 81.8 | 58.2 | 148.4 | 20.6 | 8.4 |
| *BT5-7* | 50.6 | 90.7 | 58.6 | 87.5 | 44.5 | 82.5 | 48.7 | 56.5 |
| *BT6-17* | 26.5 | 77.7 | 31.7 | 69.4 | 16.2 | 98.2 | 26.1 | 35.0 |
| *BT6-19* | 74.5 | 79.4 | 68.8 | 76.0 | 78.0 | 85.3 | 17.7 | 10.8 |
| | Percentage respect to control conditions (no water shortage) | | | | | | | |

Regarding the reduction in the root dry weight, the landscape was even more enlightening, since all G $\times$ G combinations strongly mitigate the effect of drought conditions with respect to the reduction percentages obtained for each cultivar in the absence of rhizobia (with the exception of 5-RIL33, in which the root growth was not affected by the water shortage) (Table 6, Figure 4 and Supplementary Dataset S4).

Finally, in drought-stressing conditions, the nodule development in all rhizobia–chickpea combinations was severely compromised, since both nodule parameters were strongly reduced under limited watering (Figure 4).

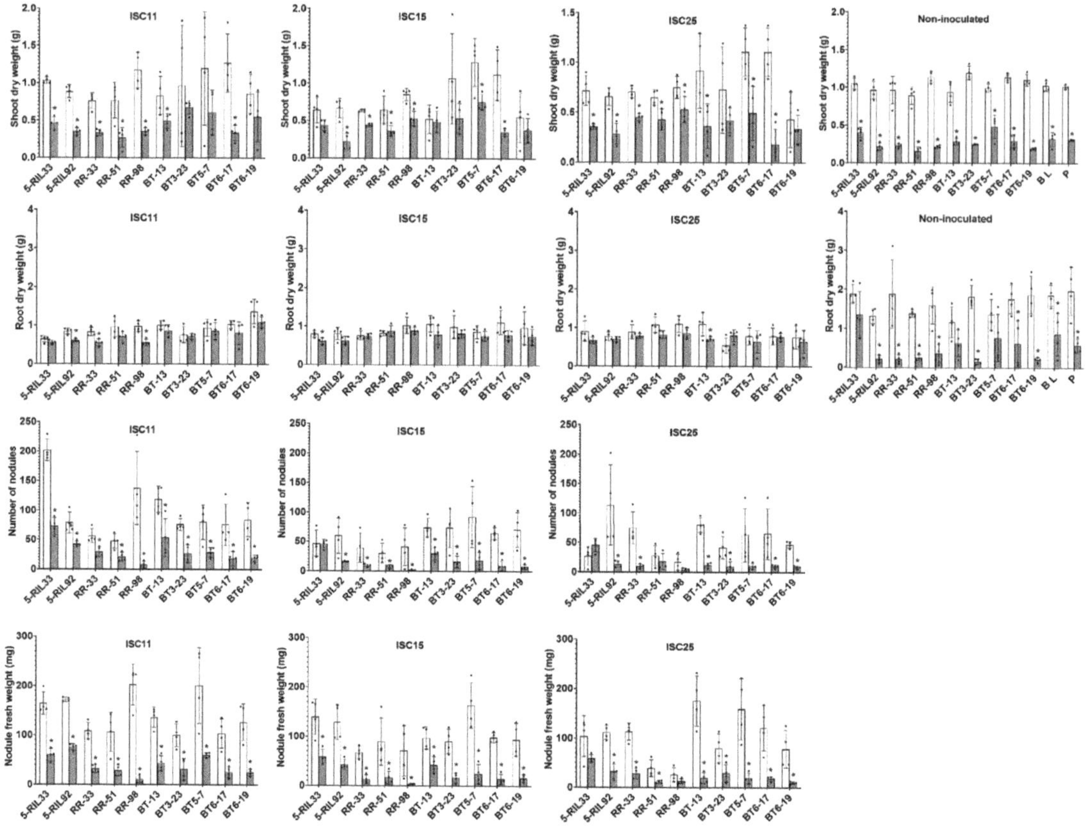

**Figure 4.** Variability in symbiotic performance among strains and chickpea cultivars. Figure shows the shoot and root dry weight and the number and fresh weight of nodules measured for each plant inoculated with the different *Mesorhizobium* strains. Control and drought conditions are indicated as white and grey bars, respectively. Multiple comparisons were calculated using *t* test ($p < 0.05$). Asterisks indicate statistically significant differences between control and drought conditions for each G × G combination.

## 4. Discussion

The present study highlights the promising potential of selected *Mesorhizobium* strains (ISC11, ISC15, and ISC25) in enhancing chickpea performance under water-limiting conditions. Our multi-level approach—spanning from phenotypic screening and greenhouse trials to genome sequencing and functional annotation—enabled the identification of elite strains with cultivar-specific symbiotic efficiency and plant growth-promoting (PGP) traits. The importance of inoculating chickpea with appropriate rhizobial strains is especially significant in regions where native rhizobia are either absent or exhibit low compatibility with target cultivars [22]. This is crucial in the Mediterranean and semi-arid areas, where environmental constraints such as drought are intensifying due to climate change. By enabling efficient biological nitrogen fixation (BNF), the use of elite *Mesorhizobium* strains not only reduces the reliance on chemical fertilizers but also contributes to long-term soil fertility, thereby promoting sustainable agricultural practices. Actually, the utilization of selected *Mesorhizobium* strains for chickpea bioinoculation offers numerous environmental and economic benefits. Indeed, it is generally considered that when chickpeas are introduced into a new region where native chickpea rhizobia are not present, it is essential to inoculate plant material with specific rhizobia to benefit from symbiotic nitrogen fixation [7]. In general, by reducing the need for synthetic nitrogen fertilizers, symbiotic nitrogen fixation operated by *Mesorhizobium* in chickpea mitigates the negative environmental impacts associated with fertilizer runoff and emissions. Moreover, it contributes to soil health by promoting nutrient cycling and improving soil structure through the formation of root nodules and finally, the reduced input costs associated with fertilizers can lead to economic savings for farmers. However, the large variability in the symbiotic performances of indigenous *Mesorhizobium* strains risks producing inconsistent yields and important differences when using a non-inoculated seed [45].

Here we have reported the isolation and characterization of three novel *Mesorhizobium* strains (ISC11, ISC15, ISC25), belonging to the species *M. ciceri*, *M. mediterraneum,* and to a possibly novel species (ISC25) closely related to *M. tianshanense*, which requires further analysis to accurately determine its true taxonomic classification. The strain ISC25 highlights the under-explored microbial diversity associated with chickpea in Mediterranean soils (see for instance [46]) which opens opportunities for future bioprospecting. These strains show important symbiotic qualities over two commercial cultivars (Blanco Lechoso and Pedrosillano) and could led to increase chickpea yield under drought conditions. Functionally, all three elite strains displayed a repertoire of plant growth-promoting traits, including genes involved in stress control, nutrient acquisition, and plant signalling. Notably, ISC11 and ISC25 harboured the *acdS* gene, encoding ACC deaminase, which is known to mitigate ethylene-mediated stress responses in plants [47] and has been demonstrated to have effects on salt stress in chickpea [48]. This enzyme likely plays a central role in enhancing drought resilience. Furthermore, the presence of quorum sensing pathways and biosynthetic gene clusters, particularly in ISC25, may indicate additional regulatory or signalling functions beneficial to plant–microbe interactions [49]. These results are in accordance with the good behaviour showed by these strains under greenhouse experiments and reinforce the idea that in silico genome analyses are a good tool for detecting potential plant growth-promoting bacteria proposed by Flores-Felix et al. [50]. We further evaluated in ten chickpea accessions if the effects of inoculation could vary depending on the combination of plant genotype (accession) × bacterial genotype (strain) (G × G) in normal and water-limiting conditions. In fact, previous studies have reported for symbiotic rhizobia large variation in the host plant × rhizobial strain combinations in plant growth [44,51]. The comprehensive testing of the strains under controlled drought conditions revealed that rhizobial inoculation not only alleviates biomass loss in shoot tissues but has a more pronounced effect on root system maintenance. Notably, in seven of the ten chickpea cultivars tested (only the 5-RIL92, BT5-7 and BT6-17 cultivars did not respond positively to the presence of the selected mesorhizobia) at least one of the combinations with the *Mesorhizobium* strains significantly reduced the impact of water shortage on shoot growth. This effect was even more drastic in the case of root development, since in all G × G combinations, the reduction in weight was completely tackled under drought conditions. This could be a consequence of the general nitrogen levels of chickpea plants, since it is well known that the status of this nutrient has a profound influence on their sensitivity to subsequent water deprivation [52,53]. In legumes, nodulated plants exhibit a marked delay in drought-induced leaf senescence compared with non-nodulated plants by means of potassium concentration increase, a better balance in carbon partitioning between starch and sugars, and the enhancement of osmolyte reserves during drought. As a result, nodulated plants generally recover more effectively from drought than their non-nodulated counterparts [54]. This resilience in root biomass could be pivotal for long-term plant survival under water deficit scenarios, as roots play a fundamental role in water and nutrient uptake. The observation that nodulation was impaired under drought, yet biomass was preserved, points to additional plant growth promotion mechanisms beyond nitrogen fixation, such as hormonal modulation or enhanced water use efficiency. From an ecological and agronomic standpoint, these findings align with the broader goals of sustainable intensification and climate adaptation in agriculture [55]. By customizing inoculant strategies based on local chickpea germplasms and specific rhizobial strains, it is possible to maximize the symbiotic benefits and minimize the environmental risks associated with conventional fertilization practices.

## 5. Conclusions

The primary long-term goal of this research was to contribute to the support and enhancement of chickpea production, which serves as a key source of plant-based proteins, in the face of an increasingly arid and challenging Mediterranean environment. This study identifies and characterizes three elite *Mesorhizobium* strains with the capacity to significantly enhance chickpea growth under drought conditions through strain-specific symbiotic interactions and plant growth-promoting mechanisms. Our integrated phenotypic and genomic approach demonstrates the following:

- ISC11 and ISC15 represent effective, well-characterized symbionts (*M. ciceri* and *M. mediterraneum*), while ISC25 may constitute a novel species with potential for bioinoculant development.
- Rhizobial performance is strongly influenced by the chickpea cultivar, reinforcing the necessity of G × G screening to optimize biofertilizer application.
- The presence of ACC deaminase and auxin-related genes in the elite strains suggests that they can mitigate abiotic stress not only via nitrogen fixation but also through hormonal balance and improved root architecture.
- The strains demonstrated potential in buffering drought effects, particularly on root biomass, supporting their utility in semi-arid and drought-prone agroecosystems.

Future research should expand to field-level trials and examine seed yield and nutrient content as ultimate indicators of agronomic performance. Additionally, the long-term monitoring of soil microbiota in inoculated versus non-inoculated systems would offer insights into the ecological compatibility and sustainability of these bioinoculants. The approach and findings outlined here provide a solid foundation for developing cultivar-specific rhizobial inoculants tailored to the challenges of climate-resilient agriculture.

**Supplementary Materials:** The following supporting information can be downloaded at: https://www.mdpi.com/article/10.3390/agriculture15151694/s1, Supplementary Materials File; Supplementary Dataset S1: Taxonomic assignment based on dDDH values; Supplementary Dataset S2: Roary summary; Supplementary Dataset S3: Results from PLaBAse ontology analysis. The number of orthologs for each ontology is reported; Supplementary Dataset S4: Data regarding the interactions between plant and rhizobia genotypes, in normal irrigation and drought conditions; Table S1: Locus tags (PGAP NCBI annotation) of genes encoding the aminocyclopropane-1-carboxylate deaminase (acdS gene); Table S2: Secondary metabolite gene clusters after search with AntiSMASH 7.0; Figure S1: Phylogeny of NodC and NifH proteins. Phylogenetic trees obtained from NodC (a) and NifH (b) protein sequences of strains ISC11, ISC15, and ISC25 are reported. NodC and NifH proteins were first identified by BLASTp. Phylogenies were estimated with the maximum likelihood method with the JTT matrix and a bootstrap of $n = 1000$ replicates (see Section 2); Figure S2: Presence of genes for the pathway related to IAA and zeatin production in the sequenced strains. Maps from tryptophan metabolism (a) and zeatin production (b) of KEGG pathways obtained by KAAS annotation for ISC11, ISC15, and ISC25 strains. Green colours indicate the presence of an orthologous gene coding for the enzyme carrying out the reaction (EC enzymes codes are indicated). Blue boxes refer to the presence of the reaction into KEGG reference.

**Author Contributions:** Conceptualization, C.F., A.M., F.M., P.B. and D.N.R.-N.; investigation, all authors; writing—original draft preparation, F.V., C.F., A.M., M.C., M.N., F.J.O. and F.P.-M.; writing—review and editing, all authors. All authors have read and agreed to the published version of the manuscript.

**Funding:** This research was funded by the European Union under the Grant Agreement no. 101102316 (LEGU-MED project, Number: 2019-SECTION2-19). The Spanish Subproject is funded by AEI/10.13039/501100011033.

**Institutional Review Board Statement:** Not applicable.

**Data Availability Statement:** Sequences are deposited in NCBI database under BioProject PRJNA951646.

**Acknowledgments:** We acknowledge the contribution of Alice Cama and Greta Soldani in technically assisting the symbiotic assays and preliminary genome analyses.

**Conflicts of Interest:** The authors declare no conflicts of interest.

## References

1. Brun, P.; Camacho, M.; Perea, F.; Rubio, M.J.; Rodríguez-Navarro, D.N. Characterization of Spanish Chickpea Genotypes (*Cicer arietinum* L.): Proximate, Mineral, and Phenolic Compounds Composition. *Eur. Food Res. Technol.* **2024**, *250*, 1007–1016. [CrossRef]
2. Goldstein, J. Faculty Opinions Recommendation of Food in the Anthropocene: The EAT-Lancet Commission on Healthy Diets from Sustainable Food Systems. *Fac. Opin.—Post-Publ. Peer Rev. Biomed. Lit.* **2019**, *393*, 447–492. [CrossRef]
3. Beebe, S.E.; Rao, I.M.; Blair, M.W.; Acosta-Gallegos, J.A. Phenotyping Common Beans for Adaptation to Drought. *Front. Physiol.* **2013**, *4*, 35. [CrossRef] [PubMed]
4. Polania, J.; Poschenrieder, C.; Rao, I.; Beebe, S. Estimation of Phenotypic Variability in Symbiotic Nitrogen Fixation Ability of Common Bean under Drought Stress Using 15N Natural Abundance in Grain. *Eur. J. Agron.* **2016**, *79*, 66–73. [CrossRef]
5. Martinelli, F.; Ollero, F.J.; Giovino, A.; Perrone, A.; Bekki, A.; Sikora, S.; Nabbout, R.E.; Bouhadida, M.; Yucel, D.; Bazzicalupo, M.; et al. Proposed Research for Innovative Solutions for Chickpeas and Beans in a Climate Change Scenario: The Mediterranean Basin. *Sustainability* **2020**, *12*, 1315. [CrossRef]
6. Cuellar-ortiz, S.M.; De La Paz Arrieta-montiel, M.; Acosta-gallegos, J.; Covarrubias, A.A. Relationship between Carbohydrate Partitioning and Drought Resistance in Common Bean. *Plant Cell Environ.* **2008**, *31*, 1399–1409. [CrossRef]
7. Zhang, J.; Wang, J.; Zhu, C.; Singh, R.P.; Chen, W. Chickpea: Its Origin, Distribution, Nutrition, Benefits, Breeding, and Symbiotic Relationship with *Mesorhizobium* Species. *Plants* **2024**, *13*, 429. [CrossRef]
8. Gunnabo, A.H.; Van Heerwaarden, J.; Geurts, R.; Wolde-Meskel, E.; Degefu, T.; Giller, K.E. Symbiotic Interactions between Chickpea (*Cicer arietinum* L.) Genotypes and Mesorhizobium Strains. *Symbiosis* **2020**, *82*, 235–248. [CrossRef]
9. Andrews, M.; Andrews, M.E. Specificity in Legume-Rhizobia Symbioses. *Int. J. Mol. Sci.* **2017**, *18*, 705. [CrossRef]
10. Zaw, M.; Rathjen, J.R.; Zhou, Y.; Ryder, M.H.; Denton, M.D. Symbiotic Effectiveness, Ecological Adaptation and Phylogenetic Diversity of Chickpea Rhizobia Isolated from a Large-Scale Australian Soil Collection. *Plant Soil* **2021**, *469*, 49–71. [CrossRef]
11. Nour, S.M.; Fernandez, M.P.; Normand, P.; Cleyet-Marel, J. *Rhizobium ciceri* sp. nov., Consisting of Strains That Nodulate Chickpeas (*Cicer arietinum* L.). *Int. J. Syst. Bacteriol.* **1994**, *44*, 511–522. [CrossRef]
12. Nour, S.M.; Cleyet-Marel, J.; Normand, P.; Fernandez, M.P. Genomic Heterogeneity of Strains Nodulating Chickpeas (*Cicer arietinum* L.) and Description of *Rhizobium mediterraneum* sp. nov. *Int. J. Syst. Bacteriol.* **1995**, *45*, 640–648. [CrossRef]
13. Zhang, J.J.; Liu, T.Y.; Chen, W.F.; Wang, E.T.; Sui, X.H.; Zhang, X.X.; Li, Y.; Li, Y.; Chen, W.X. *Mesorhizobium muleiense* sp. nov., Nodulating with *Cicer arietinum* L. *Int. J. Syst. Evol. Microbiol.* **2012**, *62*, 2737–2742. [CrossRef]
14. Zhang, J.; Guo, C.; Chen, W.; De Lajudie, P.; Zhang, Z.; Shang, Y.; Wang, E.T. *Mesorhizobium wenxiniae* sp. nov., Isolated from Chickpea (*Cicer arietinum* L.) in China. *Int. J. Syst. Evol. Microbiol.* **2018**, *68*, 1930–1936. [CrossRef] [PubMed]
15. Spaink, H.P.; Kondorosi, A.; Hooykaas, P.J.J. *The Rhizobiaceae: Molecular Biology of Model Plant-Associated Bacteria*; Springer Science & Business Media: Berlin/Heidelberg, Germany, 2012.
16. Rivas, R.; Laranjo, M.; Mateos, P.F.; Oliveira, S.; Martínez-Molina, E.; Velázquez, E. Strains of *Mesorhizobium amorphae* and *Mesorhizobium tianshanense*, Carrying Symbiotic Genes of Common Chickpea Endosymbiotic Species, Constitute a Novel Biovar (Ciceri) Capable of Nodulating *Cicer arietinum*. *Lett. Appl. Microbiol.* **2006**, *44*, 412–418. [CrossRef] [PubMed]
17. Laranjo, M.; Alexandre, A.; Oliveira, S. Legume Growth-Promoting Rhizobia: An Overview on the *Mesorhizobium* Genus. *Microbiol. Res.* **2013**, *169*, 2–17. [CrossRef] [PubMed]
18. Tena, W.; Wolde-Meskel, E.; Degefu, T.; Walley, F. Genetic and Phenotypic Diversity of Rhizobia Nodulating Chickpea (*Cicer arietinum* L.) in Soils from Southern and Central Ethiopia. *Can. J. Microbiol.* **2017**, *63*, 690–707. [CrossRef]

19. Alexandre, A.; Brígido, C.; Laranjo, M.; Rodrigues, S.; Oliveira, S. Survey of Chickpea Rhizobia Diversity in Portugal Reveals the Predominance of Species Distinct from *Mesorhizobium ciceri* and *Mesorhizobium mediterraneum*. *Microb. Ecol.* **2009**, *58*, 930–941. [CrossRef]
20. Laranjo, M.; Machado, J.; Young, J.P.W.; Oliveira, S. High Diversity of Chickpea *Mesorhizobium* Species Isolated in a Portuguese Agricultural Region. *FEMS Microbiol. Ecol.* **2004**, *48*, 101–107. [CrossRef]
21. Romdhane, S.B.; Aouani, M.E.; Mhamdi, R. Inefficient nodulation of chickpea (*Cicer arietinum* L.) in the arid and Saharan climates in Tunisia by *Sinorhizobium meliloti* biovar medicaginis. *Ann. Microbiol.* **2007**, *57*, 15–20. [CrossRef]
22. Romdhane, S.B.; De Lajudie, P.; Fuhrmann, J.J.; Mrabet, M. Potential Role of Rhizobia to Enhance Chickpea-Growth and Yield in Low Fertility-Soils of Tunisia. *Antonie Van Leeuwenhoek* **2022**, *115*, 921–932. [CrossRef]
23. Yadav, P.; Chandra, R.; Anusandhan, N.P. Plant Growth Promoting Mesorhizobia as a Potential Inoculant for Chickpea (*Cicer arietinum* L.): A Review. *Bhartiya Krishi Anusandhan Patrika* **2022**, *37*, 328–333. [CrossRef]
24. Nascimento, F.X.; Brígido, C.; Glick, B.R.; Oliveira, S. ACC Deaminase Genes Are Conserved among *Mesorhizobium* species Able to Nodulate the Same Host Plant. *FEMS Microbiol. Lett.* **2012**, *336*, 26–37. [CrossRef] [PubMed]
25. Somasegaran, P.; Hoben, H. *Methods in Legume-Rhizobium Technology*; University of Hawaii NifTAL Project and MIRCEN, Department of Agronomy and Soil Science, Hawaii Institute of Tropical Agriculture and Human Resources, College of Tropical Agriculture and Human Resources: Maui-Paia, HI, USA, 1985.
26. Beringer, J.E. R Factor Transfer in Rhizobium Leguminosarum. *Microbiology* **1974**, *84*, 188–198. [CrossRef] [PubMed]
27. Schwartz, W.J.M. Vincent, A Manual for the Practical Study of the Root-Nodule Bacteria (IBP Handbuch No. 15 Des International Biology Program, London). XI u. 164 S., 10 Abb., 17 Tab., 7 Taf. Oxford-Edinburgh 1970: Blackwell Scientific Publ., 45 s. *Z. Allg. Mikrobiol.* **1972**, *12*, 440. [CrossRef]
28. Del Cerro, P.; Rolla-Santos, A.A.P.; Gomes, D.F.; Marks, B.B.; Del Rosario Espuny, M.; Rodríguez-Carvajal, M.Á.; Soria-Díaz, M.E.; Nakatani, A.S.; Hungria, M.; Ollero, F.J.; et al. Opening the "Black Box" of nodD3, nodD4 and nodD5 Genes of Rhizobium Tropici Strain CIAT 899. *BMC Genom.* **2015**, *16*, 864. [CrossRef]
29. Del Cerro, P.; Rolla-Santos, A.A.P.; Gomes, D.F.; Marks, B.B.; Pérez-Montaño, F.; Rodríguez-Carvajal, M.Á.; Nakatani, A.S.; Gil-Serrano, A.; Megías, M.; Ollero, F.J.; et al. Regulatory nodD1 and nodD2 Genes of Rhizobium Tropici Strain CIAT 899 and Their Roles in the Early Stages of Molecular Signaling and Host-Legume Nodulation. *BMC Genom.* **2015**, *16*, 251. [CrossRef]
30. Buendia-Claveria, A.M.; Ruiz-Sainz, J.E.; Cubo-Sanchez, T.; Perez-Silva, J. Studies of Symbiotic Plasmids in *Rhizobium trifolii* and Fast-growing Bacteria That Nodulate Soybeans. *J. Appl. Microbiol.* **1986**, *61*, 1–9. [CrossRef]
31. Checcucci, A.; diCenzo, G.C.; Ghini, V.; Bazzicalupo, M.; Becker, A.; Decorosi, F.; Döhlemann, J.; Fagorzi, C.; Finan, T.M.; Fondi, M.; et al. Creation and Characterization of a Genomically Hybrid Strain in the Nitrogen-Fixing Symbiotic Bacterium *Sinorhizobium meliloti*. *ACS Synth. Biol.* **2018**, *7*, 2365–2378. [CrossRef]
32. R Core Team. *R: A Language and Environment for Statistical Computing*; R Foundation for Statistical Computing: Vienna, Austria, 2014.
33. Puppo, A.; Rigaud, J. A Possible Reduction Pathway for Leghemoglobin in Vivo. *FEBS Lett.* **1979**, *108*, 124–126. [CrossRef]
34. Tatusova, T.; DiCuccio, M.; Badretdin, A.; Chetvernin, V.; Nawrocki, E.P.; Zaslavsky, L.; Lomsadze, A.; Pruitt, K.D.; Borodovsky, M.; Ostell, J. NCBI Prokaryotic Genome Annotation Pipeline. *Nucleic Acids Res.* **2016**, *44*, 6614–6624. [CrossRef]
35. Meier-Kolthoff, J.P.; Göker, M. TYGS Is an Automated High-Throughput Platform for State-of-the-Art Genome-Based Taxonomy. *Nat. Commun.* **2019**, *10*, 2182. [CrossRef]
36. Moriya, Y.; Itoh, M.; Okuda, S.; Yoshizawa, A.C.; Kanehisa, M. KAAS: An Automatic Genome Annotation and Pathway Reconstruction Server. *Nucleic Acids Res.* **2007**, *35*, W182–W185. [CrossRef]
37. Blin, K.; Shaw, S.; Augustijn, H.E.; Reitz, Z.L.; Biermann, F.; Alanjary, M.; Fetter, A.; Terlouw, B.R.; Metcalf, W.W.; Helfrich, E.J.N.; et al. AntiSMASH 7.0: New and Improved Predictions for Detection, Regulation, Chemical Structures and Visualisation. *Nucleic Acids Res.* **2023**, *51*, W46–W50. [CrossRef]
38. Patz, S.; Gautam, A.; Becker, M.; Ruppel, S.; Rodríguez-Palenzuela, P.; Huson, D.H. PLaBAse: A Comprehensive Web Resource for Analyzing the Plant Growth-Promoting Potential of Plant-Associated Bacteria. *bioRxiv* **2021**. bioRxiv:2021.12.13.472471. [CrossRef]
39. Seemann, T. Prokka: Rapid Prokaryotic Genome Annotation. *Bioinformatics* **2014**, *30*, 2068–2069. [CrossRef] [PubMed]
40. Page, A.J.; Cummins, C.A.; Hunt, M.; Wong, V.K.; Reuter, S.; Holden, M.T.G.; Fookes, M.; Falush, D.; Keane, J.A.; Parkhill, J. Roary: Rapid Large-Scale Prokaryote Pan Genome Analysis. *Bioinformatics* **2015**, *31*, 3691–3693. [CrossRef] [PubMed]
41. Tamura, K.; Stecher, G.; Kumar, S. MEGA11: Molecular Evolutionary Genetics Analysis Version 11. *Mol. Biol. Evol.* **2021**, *38*, 3022–3027. [CrossRef]
42. Jones, D.T.; Taylor, W.R.; Thornton, J.M. The Rapid Generation of Mutation Data Matrices from Protein Sequences. *Bioinformatics* **1992**, *8*, 275–282. [CrossRef] [PubMed]
43. Lefort, V.; Desper, R.; Gascuel, O. FastME 2.0: A Comprehensive, Accurate, and Fast Distance-Based Phylogeny Inference Program. *Mol. Biol. Evol.* **2015**, *32*, 2798–2800. [CrossRef]

44. Fagorzi, C.; Bacci, G.; Huang, R.; Cangioli, L.; Checcucci, A.; Fini, M.; Perrin, E.; Natali, C.; diCenzo, G.C.; Mengoni, A. Nonadditive Transcriptomic Signatures of Genotype-by-Genotype Interactions during the Initiation of Plant-Rhizobium Symbiosis. *mSystems* **2021**, *6*. [CrossRef]
45. Kyei-Boahen, S.; Slinkard, A.E.; Walley, F.L. Evaluation of Rhizobial Inoculation Methods for Chickpea. *Agron. J.* **2002**, *94*, 851–859. [CrossRef]
46. Romdhane, S.B.; Trabelsi, M.; Aouani, M.E.; De Lajudie, P.; Mhamdi, R. The Diversity of Rhizobia Nodulating Chickpea (*Cicer arietinum*) under Water Deficiency as a Source of More Efficient Inoculants. *Soil Biol. Biochem.* **2009**, *41*, 2568–2572. [CrossRef]
47. Singh, R.P.; Shelke, G.M.; Kumar, A.; Jha, P.N. Biochemistry and Genetics of ACC Deaminase: A Weapon to "Stress Ethylene" Produced in Plants. *Front. Microbiol.* **2015**, *6*, 937. [CrossRef]
48. Br Igido, C.; Nascimento, F.X.; Duan, J.; Glick, B.R.; Oliveira, S. Expression of an Exogenous 1-Aminocyclopropane-1-Carboxylate Deaminase Gene in *Mesorhizobium* spp. Reduces the Negative Effects of Salt Stress in Chickpea. *FEMS Microbiol. Lett.* **2013**, *349*, 46–53. [CrossRef]
49. Poppeliers, S.W.; Sánchez-Gil, J.J.; de Jonge, R. Microbes to Support Plant Health: Understanding Bioinoculant Success in Complex Conditions. *Curr. Opin. Microbiol.* **2023**, *73*, 102286. [CrossRef]
50. Flores-Félix, J.D.; Velázquez, E.; Martínez-Molina, E.; González-Andrés, F.; Squartini, A.; Rivas, R. Connecting the Lab and the Field: Genome Analysis of Phyllobacterium and Rhizobium Strains and Field Performance on Two Vegetable Crops. *Agronomy* **2021**, *11*, 1124. [CrossRef]
51. Cangioli, L.; Checcucci, A.; Mengoni, A.; Fagorzi, C. Legume Tasters: Symbiotic Rhizobia Host Preference and Smart Inoculant Formulations. *Biol. Commun.* **2021**, *66*, 47–54. [CrossRef]
52. Radin, J.W.; Ackerson, R.C. Water Relations of Cotton Plants under Nitrogen Deficiency. *Plant Physiol.* **1981**, *67*, 115–119. [CrossRef]
53. Radin, J.W.; Boyer, J.S. Control of Leaf Expansion by Nitrogen Nutrition in Sunflower Plants: Role of Hydraulic Conductivity and Turgor. *Plant Physiol.* **1982**, *69*, 771–775. [CrossRef]
54. Staudinger, C.; Mehmeti-Tershani, V.; Gil-Quintana, E.; Gonzalez, E.M.; Hofhansl, F.; Bachmann, G.; Wienkoop, S. Evidence for a Rhizobia-Induced Drought Stress Response Strategy in *Medicago truncatula*. *J. Proteom.* **2016**, *136*, 202–213. [CrossRef]
55. Pretty, J. Intensification for Redesigned and Sustainable Agricultural Systems. *Science (1979)* **2018**, *362*. [CrossRef]

agriculture

MDPI

Article

# Microbial Population in *Curcuma* Species at Different Growth Stages

**Neptu Islamy Raharja [1], Mohammad Amzad Hossain [1,2,*] and Hikaru Akamine [1,2]**

[1] The United Graduate School of Agricultural Sciences, Kagoshima University, Kagoshima 890-0065, Japan; neptuislamy@gmail.com (N.I.R.); akamineh@agr.u-ryukyu.ac.jp (H.A.)
[2] Faculty of Agriculture, University of The Ryukyus, Nishihara, Okinawa 903-0213, Japan
* Correspondence: amzad@agr.u-ryukyu.ac.jp; Tel.: +81-98-895-8817

**Abstract:** Turmeric (*Curcuma* spp.) is widely cultivated in tropical regions for its use in traditional medicine and culinary purposes. This study investigated the bacterial populations in the rhizosphere, stems, and leaves of the *Curcuma* species and strains at different growth stages. Bacterial population cultivated in the field and plastic house showed variations across growth stages. The rhizosphere possessed the highest bacterial populations in both experiments (1.8 to 11.9 $\times 10^6$ CFU/g and 1.7 to 24.3 $\times 10^6$ CFU/g, respectively), with *C. amada* and Ryudai gold as the highest. Endophytic bacteria in stems and leaves also peaked at the middle growth stage. Principal Component Analysis (PCA) revealed distinct separations among *Curcuma* species planted in the field and plastic house at different growth stages. *C. aromatica* and *C. longa* strain L2 clustered differently under field conditions, while *C. zedoaria* and *C. xanthorrhiza* were distinct under plastic house conditions. Combined PCA revealed a clear separation between the field and plastic house, with tighter clustering observed in the plastic house. Leaf-associated bacterial populations were compositionally distinct from those in the rhizosphere and stems. These findings suggest that the *Curcuma* growth stage and species significantly affect bacterial community structure, supporting the development of targeted cultivation strategies and microbial applications to enhance productivity and sustainability in turmeric farming.

**Keywords:** turmeric; endophyte bacteria; plant growth bacteria; rhizosphere bacteria; different stages

## 1. Introduction

*Curcuma*, a diverse genus of Zingiberaceae, encompasses several species of significant economic, medicinal, and agricultural importance [1]. Among them, *Curcuma longa* L., commonly known as turmeric, is highly valued for its culinary applications and bioactive compounds, particularly curcumin [2]. This polyphenolic compound is renowned for its antioxidant, anti-inflammatory, and antimicrobial properties, making turmeric a vital component in traditional medicine, functional foods, and pharmaceutical research [2,3]. The widespread cultivation of *Curcuma* species, particularly in tropical and subtropical regions, underscores the need for a deeper understanding of the biological and ecological factors influencing their growth and productivity. One of the most critical yet often overlooked factors in plant development is the role of microbial communities associated with plant roots, leaves, and internal tissues [4]. These microorganisms contribute to plant health, nutrient cycling, disease resistance, and overall productivity [4,5]. In *Curcuma* species, the microbial populations present in the rhizosphere, phyllosphere, and endosphere undergo significant shifts across different growth stages, impacting plant physiology and resilience [6]. However, despite their importance, the dynamic interactions between *Curcuma* plants and their associated microbes throughout the growth cycle remain largely unexplored.

The rhizosphere, a biologically active soil zone around roots, is a key site of plant–microbe interaction [7]. Root exudates, including sugars, amino acids, and secondary metabolites, attract beneficial microbes and deter pathogens. During early growth, these microbes enhance seedling vigor, nutrient uptake, and disease suppression, laying the foundation for healthy root development and improved plant performance [6–8].

As *Curcuma* plants progress into the vegetative growth stage, characterized by rapid foliage expansion and root elongation, microbial populations in the rhizosphere shift in response to changing root exudation patterns and nutrient demands [6]. Beneficial microorganisms such as plant growth-promoting rhizobacteria (PGPR) and mycorrhizal fungi contribute to enhanced nutrient availability, improved soil structure, and increased resistance to environmental stressors [9]. These microbial communities assist in nitrogen fixation, phosphate solubilization, and the production of plant hormones such as auxins and cytokinins, all of which are essential for robust vegetative growth [9,10]. Beyond the rhizosphere, the endosphere represents a specialized niche within plant tissues where endophytic microorganisms reside. These microbes establish symbiotic relationships with their host plants, contributing to nutrient uptake, hormone regulation, and stress tolerance [7,11]. Previous studies have reported that endophytic bacteria in *Curcuma* species may enhance curcumin biosynthesis, exhibit antimicrobial activity, and serve as biocontrol agents against phytopathogens [12]. Exploring the functional roles of endophytic microbes at different growth stages offers valuable insights into plant–microbe interactions and their potential for sustainable agriculture. The phyllosphere microbiota, though often overlooked, also supports plant health by suppressing pathogens through competitive exclusion and antimicrobial production [13]. Environmental factors, such as humidity, temperature, and UV exposure, shape these microbial communities, influencing plant development throughout the *Curcuma* growth cycle [6,14]. Additionally, cultivation conditions (e.g., field vs. plastic house) can significantly alter microbial diversity and abundance [15,16], highlighting the importance of environmental context in optimizing *Curcuma* cultivation for better health and productivity.

Previous studies have predominantly focused on a single *Curcuma* species to analyze microbial diversity or identify specific functional microbial groups. It is believed that bacterial populations and their functions vary depending on plant species, plant growth stages, plant parts, and cultivation places. In response, this study introduces a novel approach by quantifying and comparing total bacterial populations across different growth stages and plant tissues (rhizosphere, stem, and leaves) in multiple *Curcuma* species cultivated in the field and plastic house. By elucidating the temporal shifts in microbial communities from germination to maturity and comparing microbial dynamics between plant species, this research seeks to advance our understanding of plant–microbe interactions. Understanding these microbial dynamics will provide valuable insights into optimizing microbial populations and functions as biofertilizer and stimulant, improving yields, and promoting environmentally sustainable farming systems for species. These findings also offer foundational information that may support future microbiome research related to plant cultivation strategies. To the best of our knowledge, this study is one of the first to compare microbial load dynamics across different growth stages in multiple *Curcuma* species cultivated in both a field and plastic house.

## 2. Materials and Methods

### 2.1. *Curcuma* Species or Strains

This study used five turmeric species for plastic house experiments: *Curcuma longa* (cultivar Ryudai gold and five strains), *C. zedoaria* (ZE), *C. xanthorrhiza* (ZA), *C. aromatica* (AR), and *C. amada* (AM). The cultivar Ryudai gold (RG) and five strains belong to the *Curcuma longa* species; the five strains are known as *C. longa* strain 1 (L1), *C. longa* strain 2 (L2), *C. longa* strain 3 (L3), *C. longa* strain 4 (L4), and *C. longa* strain 5. Four turmeric species were used in the field experiment: *Curcuma longa* (cultivar Ryudai gold, *C. longa strain* 1 (L1), *C. longa* strain 2 (L2), *C. longa* strain 3 (L3), *C. zedoaria* (ZE), *C. xanthorrhiza* (ZA), and *C. aromatica* (AR). The various species, cultivars, and strains of turmeric differ in rhizome dimensions, shapes, and colors. They also possess distinct chemical properties, flavors, tastes, and physiological as well as morphological traits of the shoot (data not yet published). Given their yield performance, these turmeric varieties have the potential for commercial production.

### 2.2. *Curcuma* Cultivation

The pot experiment was conducted in a plastic house from 2 May 2022 to 6 February 2023. Air-dried dark-red soil of 3.5 kg and cultured soil (commercial name: Hanasakimonogatari) of 2.5 kg were mixed properly and placed in each Wagner pot (0.05 $m^2$). As the rhizome sizes were different from the *Curcuma* species and strains, the best seed-rhizomes were selected for each species and strains. One seed-rhizome per pot was planted at the depth of 6 cm [17]. The pots were placed in the house randomly. The outdoor environment was maintained in the house by keeping the windows opened, but the windows were closed during typhoons. Water was applied regularly as required to maintain the optimum soil moisture level for proper seedling emergence and plant growth.

The field experiment was conducted from May 2022 to February 2023 at the Subtropical Field Science Center of the University of the Ryukyus Okinawa. Th experiment was conducted on fields of dark red soil (Shimajiri Maji, Chromic Luvisol) [18].

### 2.3. Sample Collection

We collected both the plant (leaf and stem) and rhizosphere samples three times at the early growth stage (3- to 4-leaf stage, July), the middle growth stage (7- to 9-leaf stage, October), and the late growth stage (plant maturing stage, December) when the plants were still green. The plants were cut at the soil surface, and the leaves were separated from the stems. The rhizosphere soil was collected from each plant, and a composite soil sample was prepared for each *Curcuma* species or strains.

### 2.4. Isolation and Enumeration of Endophytic Bacteria

Plant samples (stems and leaves) were cleaned by washing in running water and cut into pieces. The surface of the sample pieces was washed and sterilized to free from microbes as follows. The leaves or stems were washed with sterile distilled water and then with 70% ethanol solution for one minute. The leaves or stems were then washed with 3% sodium hypochlorite solution (Nacalai Tesque, Kyoto, Japan) for 1.5 min for leaves and 3 min for stems. Finally, they were rinsed with sterile distilled water three times [19]. The samples were then dried using sterile tissue paper. Surface sterilization of the samples were performed by spreading 0.1 mL of distilled water on Nutrient Agar media (Difco; BD, Franklin Lakes, NJ, USA), then the petri dishes were incubated (Sanyo MIR-152; Sanyo, Osaka, Japan) at 28 °C for 14 days, and it was ensured that no colonies appeared. Surface sterilized leaves or stems were crushed using a sterile mortar and pestle. The crushed sample was put into a test tube containing 9 mL of sterile physiological NaCl solution

(0.85%) then serially diluted, and 0.1 mL of each dilution was plated on a Nutrient Agar media that had been added with 50 mg/L nystatin (Nacalai Tesque, Kyoto, Japan) to inhibit the growth of the fungus [19]. The diluted samples were then plated onto petri dishes and incubated at 28 °C for 14 days. Colonies that appeared on petri dishes were counted every day for 14 days. All analyses were conducted in triplicate.

### 2.5. Isolation and Enumeration of Rhizosphere Heterotrophic Bacteria

Rhizosphere soil samples were taken by removing the *Curcuma* plant from the soil and carefully shaking the rhizome to remove loose and non-adherent soil. The soil that was still attached to the roots and rhizome was then collected using a sterile spatula, and the rhizosphere soil obtained was then put into a sterile plastic bag. Each soil composite sample of 10 g was taken and then put into 90 mL of physiological NaCl solution (0.85%). It was homogenized using an orbital shaker for 30 min at a speed of 150 rpm, followed by serial dilutions. A total of 0.1 mL of soil suspension from each dilution was spread on Nutrient Agar media. The diluted samples were then plated onto petri dishes and incubated at 28 °C for 7 days using a Sanyo MIR-152 incubator. Bacterial colonies that appeared were counted every day. All analyses were conducted in triplicate.

### 2.6. Data Analysis

The data of this study were analyzed qualitatively and quantitatively. The data were expressed as mean ± standard deviation (SD). Statistical analysis was conducted using the one-way Analysis of Variance (ANOVA) then followed up with the Tukey's test using the IBM SPSS (Statistical Program Software System) program version 26.0 (IBM Corp., Armonk, NY, USA). Significant differences were those in which $p < 0.05$. Principal component analysis (PCA) was constructed using SIMCA Version 17 (Sartorius, Gottingen, Germany).

## 3. Results

### 3.1. Effect of Different Curcuma Growth Stages on Rhizosphere Heterotrophic Bacteria Population

The bacterial population in the rhizosphere soil of *Curcuma* species or strains varied across growth stages in both experiments (Figure 1). In field-grown samples, the bacterial counts ranged from 1.8 to 9.1 × $10^6$ CFU/g during the early stage, 2.0 to 11.9 × $10^6$ CFU/g in the middle stage, and 1.9 to 10.7 × $10^6$ CFU/g at the late stage. In contrast, rhizosphere soil from *Curcuma* cultivated in a plastic house exhibited higher bacterial populations, ranging from 1.7 to 21.8 × $10^6$ CFU/g in the early stage, 1.8 to 24.3 × $10^6$ CFU/g in the middle stage, and 1.7 to 23.1 × $10^6$ CFU/g in the late stage.

Significant differences were observed in the bacterial populations depending on the *Curcuma* species or strain and their growth stage. The highest bacterial population was recorded in the rhizosphere soil of the RG turmeric strain grown in the field during the middle growth stage (11.9 × $10^6$ CFU/g) ($p < 0.05$), followed by *C. amada* cultivated in the plastic house at the same stage (24.3 × $10^6$ CFU/g).

### 3.2. Effect of the Different Curcuma Growth Stages on Endophytic Bacteria Population in the Stems

The population of endophytic bacteria in *Curcuma* stems varied significantly across growth stages and among species or strains (Figure 2). In field-grown plants, stem bacterial counts ranged from 2.0 to 9.3 × $10^5$ CFU/g, while those cultivated in a plastic house ranged from 3.2 to 15.0 × $10^5$ CFU/g. The RG strain exhibited the highest bacterial population during the middle growth stage, reaching 9.3 × $10^5$ CFU/g in the field and 15.0 × $10^5$ CFU/g in the plastic house, followed by *C. amada*.

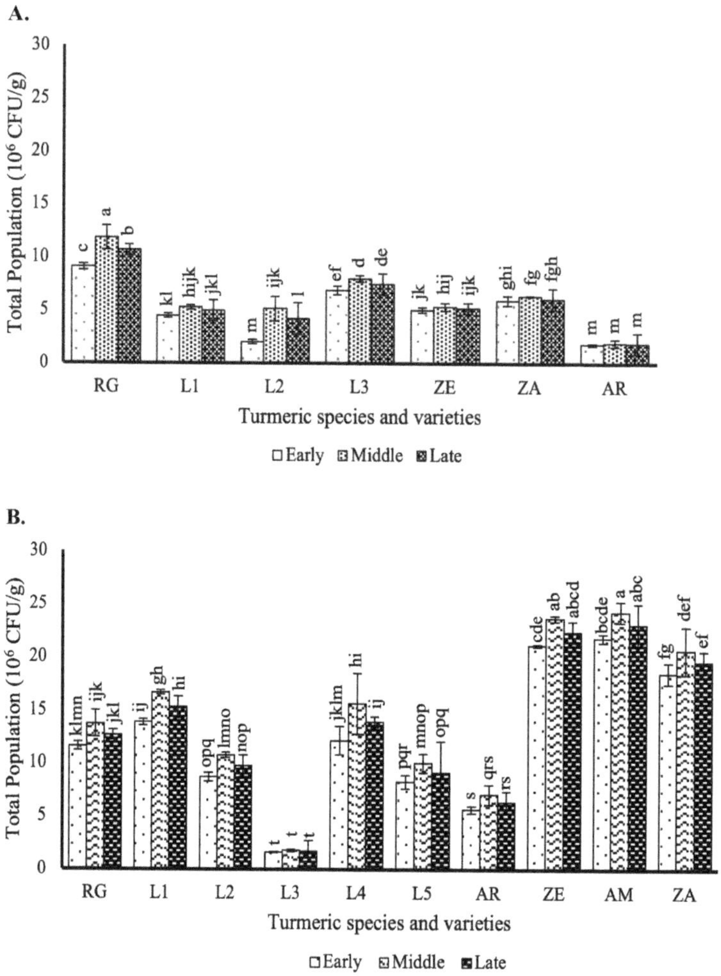


**Figure 1.** Bacterial population in the rhizosphere soil of *Curcuma* cultivated in field (**A**) and plastic house (**B**). The different letters above the bars represent that the data were significantly different based on the Tukey's test $p < 0.05$. Note: RG, Ryudai gold; L1, *C. longa* strain 1; L2, *C. longa* strain 2; L3, *C. longa* strain 3; L4, *C. longa* strain 4; L5, *C. longa* strain 5; AR, *C. aromatica*; ZE, *C. zedoaria*; *AM*, *C. amada*; ZA, *C. xanthorrhiza*.

Notably, endophytic bacterial populations differed between the two experiments for each *Curcuma* species or strain. Plants at the middle growth stage consistently showed significantly higher bacterial counts compared to other stages in both experiments ($p < 0.05$).

### 3.3. *Effect of the Different Curcuma Growth Stages on Endophytic Bacteria Population in the Leaves*

The population of endophytic bacteria in the leaves of 16 *Curcuma* species or strains varied depending on the species/strain and growth stage in both the experiments (Figure 3). In field-grown plants, bacterial populations ranged from 30.0 to 76.0 × $10^4$ CFU/g, while in plastic house-grown plants, the range was lower, from 7.3 to 23.3 × $10^4$ CFU/g.

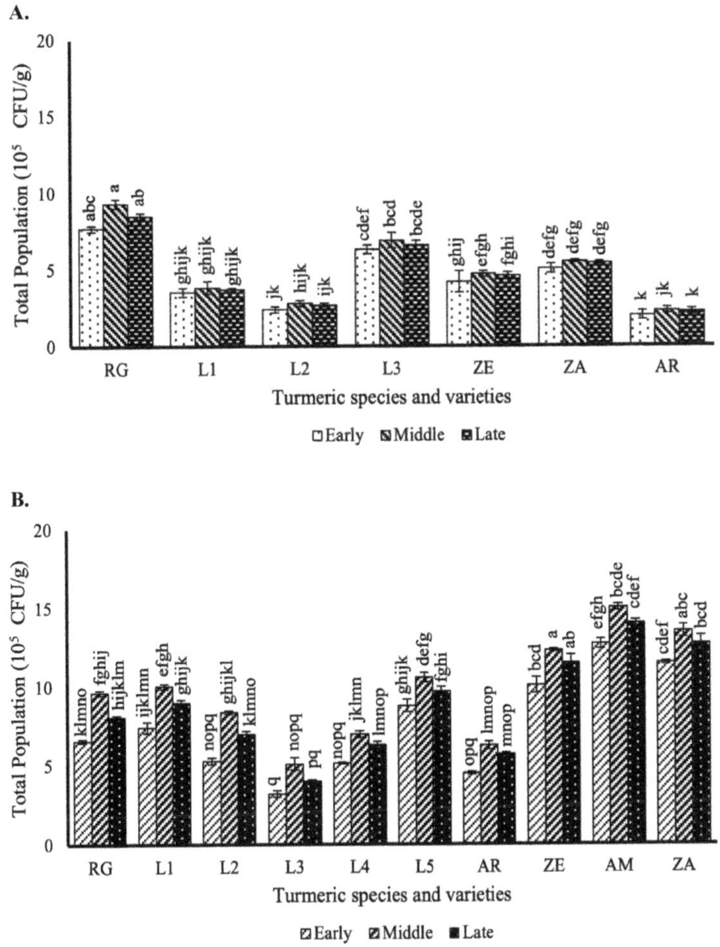


**Figure 2.** Bacterial population in the stems of *Curcuma* spp. or strains cultivated in field (**A**) and plastic house (**B**). The different letters above the bars represent that the data were significantly different based on the Tukey's test at $p < 0.05$. Note: RG, Ryudai gold; L1, *C. longa* strain 1; L2, *C. longa* strain 2; L3, *C. longa* strain 3; L4, *C. longa* strain 4; L5, *C. longa* strain 5; AR, *C. aromatica*; ZE, *C. zedoaria*; AM, *C. amada*; ZA, *C. xanthorrhiza*.

Significant differences ($p < 0.05$) were observed among growth stages, with the lowest bacterial counts recorded during the early stage and the highest during the middle stage. The *Curcuma* strain RG exhibited the highest bacterial population in both experiments ($76.0 \times 10^4$ CFU/g in the field and $23.3 \times 10^4$ CFU/g in the plastic house).

### *3.4. Multivariate Statistical Plot of the Bacterial Population in Curcuma Species Cultivated in the Field*

Principal component analysis (PCA) of bacterial populations showed the variations across seven *Curcuma* species at different growth stages. PC1 and PC2 explain 94.1% and 3.1% of the total variance, respectively. Distinct clustering patterns are observed, with *C. aromatica* and *C. longa* strain L2 clearly separated from other species along the PC1 axis, indicating differences in bacterial community composition. Other species, including *C. longa* L1 and L3, *C. zedoaria*, and *C. xanthorrhiza*, are more tightly grouped, suggesting similar microbial profiles. The variations in bacterial populations in different growth stages of *Curcuma* species were further explained by multivariate analysis using the principal component analysis (PCA). The first two principal components (PCs) explained 97.2% of the total variance, accounted for by PC1 and PC2 (94.1 and 3.12%, respectively), showing that most of the separation between *Curcuma* species occurs along the PC1 axis. The PCA analysis revealed diverse grouping patterns for bacterial population data in the rhizosphere, leaves, and stems of seven species of *Curcuma* at different growth stages. *C. aromatica* seemed to cluster in the negative direction of PC1 and PC2, indicating distinct bacterial growth patterns compared to other species (Figure 4a). Similarly, *C. longa* strain L2 displayed a distinct clustering pattern, suggesting it harbors a bacterial population different from the other *C. longa* strain L1 and L3. In contrast, *C. longa* strain L1, *C. zedoaria*, *C. xanthorrhiza*, *C. longa* strain L3, and *C. longa* strain RG were more tightly grouped, suggesting similarities in their bacterial population trends.

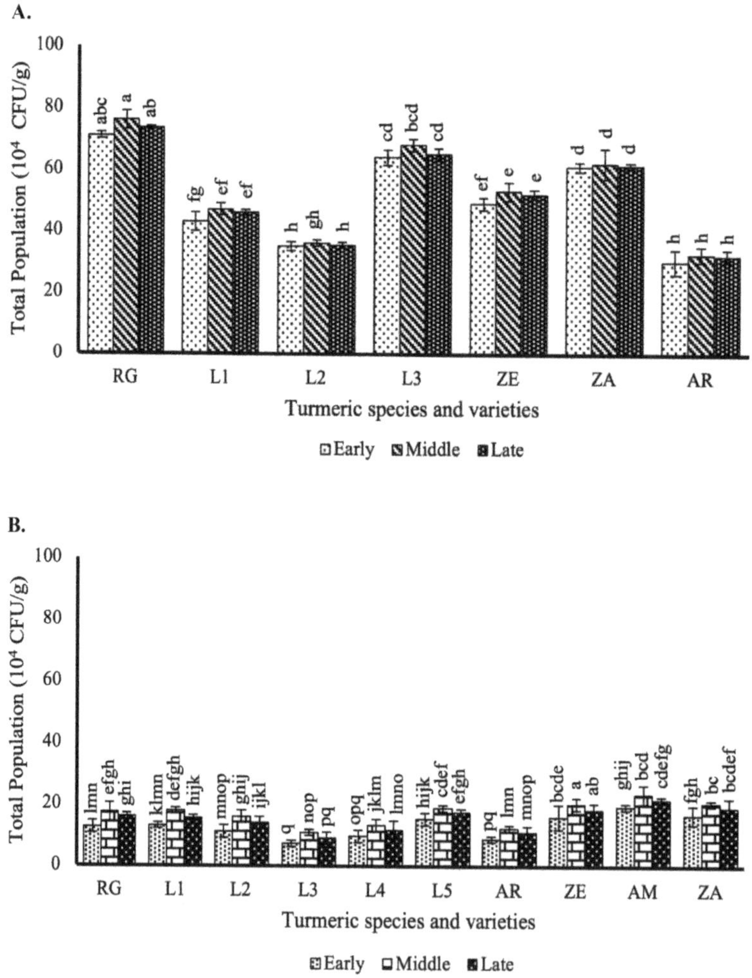


**Figure 3.** Bacterial population in the leaves of *Curcuma* spp. or strains cultivated in field (**A**) and plastic house (**B**). The different letters above the bars represent that the data were significantly different based on the Tukey's test at $p < 0.05$. Note: RG, Ryudai gold; L1, *C. longa* strain 1; L2, *C. longa* strain 2; L3, *C. longa* strain 3; L4, *C. longa* strain 4; L5, *C. longa* strain 5; AR, *C. aromatica*; ZE, *C. zedoaria*; AM, *C. amada*; ZA, *C. xanthorrhiza*.

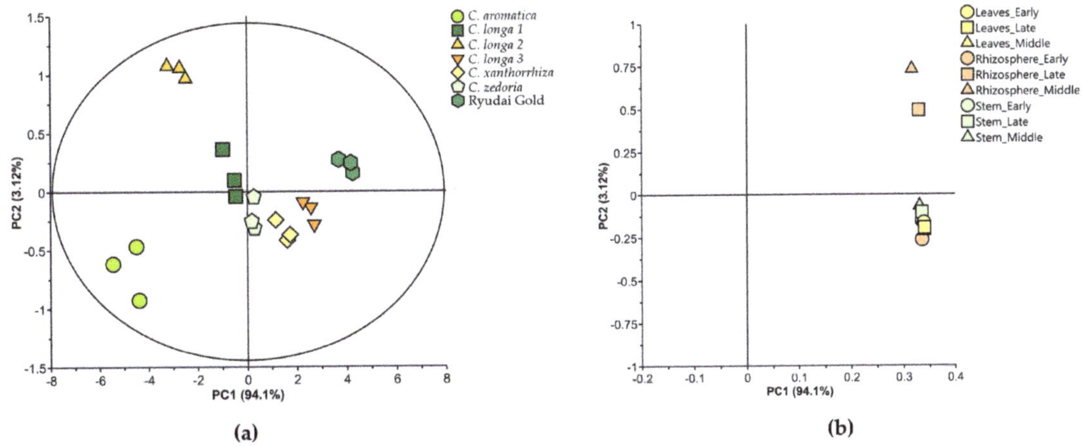


**Figure 4.** PCA of bacterial population data from different *Curcuma* species or strains and growth stages cultivated in the field. (**a**) Score plot showing sample distribution. (**b**) Loading plot indicating the contribution of variables to the principal components.

Factor loading plots were used to investigate the relationships between the score plots and the microbial growth stages variables, providing insights into how each variable contributed to the observed clustering patterns. The corresponding factor loading plot (Figure 4b) revealed that the clustering observed along PC1 was primarily driven by sample type (rhizosphere, stem, and leaf), with rhizosphere samples showing the greatest contribution to PC1 variance. The growth stage also influenced the clustering along PC2, particularly in leaf and stem samples, which exhibited greater overlap, suggesting less variation in bacterial populations across growth stages in these tissues compared to the rhizosphere. These patterns highlight that sample type was the primary driver of bacterial community differentiation, while growth stage provided additional but lesser separation, particularly in non-rhizosphere tissues. Together, these results emphasize that both plant part and growth stage play key roles in shaping microbial communities in *Curcuma*, with the rhizosphere being the most variable and influential compartment in driving overall bacterial community structure.

### *3.5. Multivariate Statistical Plot of Bacterial Population of Curcuma Species Cultivated in the Plastic House*

The first two PCs of the variations in bacterial populations in different growth stages of *Curcuma* species that were planted in the plastic house explained 97.8% of the total variance, accounted for by PC1 and PC2 (91.4 and 6.38%, respectively). *C. longa* strain L2, *C. aromatica, C. longa* strain L3, *C. longa* strain RG, and *C. longa* strain L1 form a relatively close cluster along PC1 and PC2 (Figure 5a). This indicates that these species share greater similarities in their bacterial population or growth data under plastic house conditions compared to others, such as *C. zedoaria or C. xanthorrhiza* that were plotted both in positive direction of PC1 and negative direction of PC2. Conversely, *C. longa* strain L5 exhibited a distinct separation in the score plot near to the origin (PC1 = 0, PC2 = 0), suggesting that this species has average or neutral characteristics in relation to the principal components. Thus, this indicates that its bacterial or growth data do not exhibit extreme variations and may fall somewhere in between the other species.

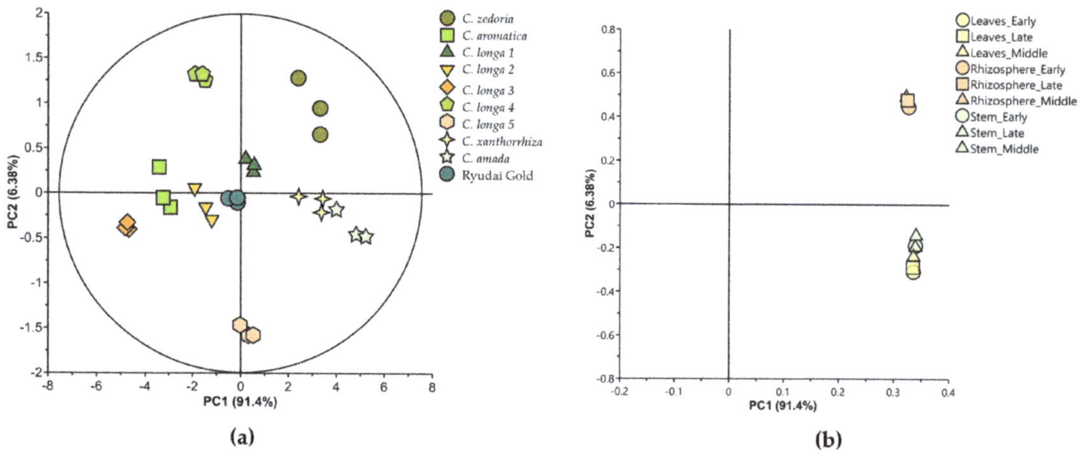


**Figure 5.** PCA of bacterial population data from Curcuma species or strains and growth stages cultivated in a plastic house. (**a**) Score plot showing sample grouping. (**b**) Loading plot representing the influence of each variable on the principal components.

The factor loading plot (Figure 5b) revealed that the primary variable contributing to separation along PC1 was the bacterial population in the rhizosphere, especially at later growth stages. These variables had high positive loadings on PC1, indicating they drive separation of species like *C. amada*, which showed dominant rhizosphere bacterial populations. In contrast, bacterial populations in the leaves and stems—particularly of *C. zedoaria* and *C. amada*, contributed more to PC2 and were associated with separation in that dimension. This suggests that while PC1 reflects rhizosphere-driven variance, PC2 captures differences in endophytic bacterial populations in aerial parts. In conclusion, rhizosphere populations dominate the separation along PC1, while endophytic bacterial differences drive PC2 variability.

### 3.6. Multivariate Statistical Plot of Combined Data on Bacterial Population in Curcuma Species Cultivated in the Field and Plastic House

The comparison between the bacterial population structure based on the growth stages of *Curcuma* species in the field and plastic house experiments was further explored using PCA, which revealed that there were distinct separations between field and plastic house (Figure 6a). Accordingly, the first two PCs further explained 95.3% of the variance and were accounted for by PC1 and PC2 (64.1 and 31.2%, respectively). Based on the score plot, the *Curcuma* species that were cultivated in the field clustered distinctly from the plastic house along PC1, indicating that the bacterial populations in the field experiment differ significantly from those in the plastic house. Moreover, the plastic house samples tend to cluster more tightly, suggesting more uniform bacterial populations, likely influenced by conditions in the plastic house. The separation of field and plastic house samples along PC1 highlights the impact of the planting environment on bacterial community structure.

The overlap between the rhizosphere and stem in the loading plot, and the separation of leaves, can be explained by differences in the bacterial community composition and diversity driven by the plant part and its microenvironment. The loading plot (Figure 6b) further clarified the contribution of specific plant parts to the observed separation. Bacterial populations associated with leaves contributed predominantly to PC2 and clustered separately from the rhizosphere and stem samples, which partially overlapped. This suggests that leaf-associated bacterial communities are compositionally distinct, while rhizosphere and stem communities are more similar, likely due to closer spatial or physiological interactions. Overall, these results emphasize that different growth stage influences the cultivation places (field vs. plastic house) as the strongest driver of community separation (PC1), while plant part, particularly leaf versus non-leaf tissues, drives secondary variation (PC2). These findings underline the influence of both external growing conditions and internal plant microenvironments in shaping microbial community structures in *Curcuma* species.

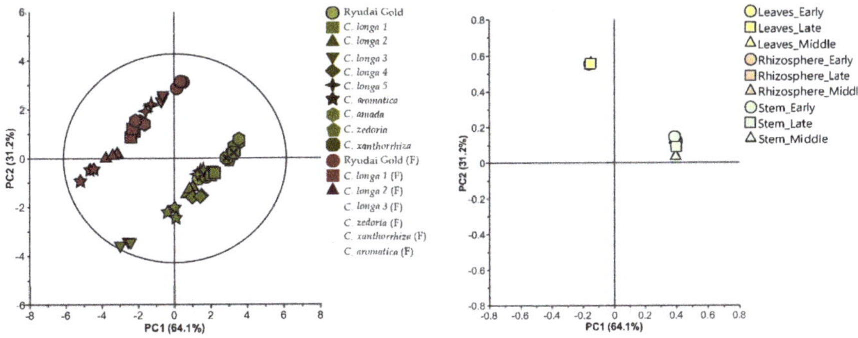


**Figure 6.** PCA based on combined bacterial population data from field and plastic house cultivation. (**a**) Score plot displaying overall sample distribution. (**b**) Loading plot illustrating the variable contributions to the principal components.

## 4. Discussion

The bacterial populations in the rhizosphere, stems, and leaves of *Curcuma* species varied significantly across different growth stages in both the experiments, highlighting the influence of plant development and external conditions on microbial communities.

Across all growth stages, rhizosphere bacterial populations were consistently higher under plastic house conditions compared to the field. This trend likely reflects the benefits of stable environmental parameters, such as temperature, humidity, and soil moisture which promote microbial proliferation [20]. This observation aligns with previous studies showing higher microbial abundance under controlled environments due to reduced abiotic stress [21]. Notably, peak bacterial density occurred at the middle growth stage in both environments, coinciding with heightened root exudation, a key factor known to enrich microbial communities by supplying carbon-rich substrates [22,23]. Moreover, root anatomical changes such as increased vascular tissue and moisture content may further enhance microbial colonization [24]. These conditions collectively enhance microbial proliferation across the rhizosphere, stem, and leaf tissues [25]. The variations observed among different *Curcuma* species and strains may be due to differences in the quantity and composition of these exudates, as plants selectively shape their rhizosphere microbiome based on their metabolic profile [26]. This is evident in the significantly higher bacterial population in Ryudai Gold and *C. amada*, suggesting that these strains may release a more favorable set of exudates for microbial colonization. In this study, the rhizosphere bacterial density was $10^6$ CFU/g, aligning with previous reports of rhizosphere populations ranging from $10^6$ to $10^9$ CFU/g [27]. The relatively low bacterial count observed may be influenced by the presence of bioactive compounds in turmeric, such as curcumin and essential oils, which are known to exhibit antimicrobial properties [28]. These compounds could potentially inhibit certain microbial populations in the rhizosphere, leading to lower overall bacterial density. Similarly, a previous study indicated that the population density of rhizosphere bacteria associated with *Zea mays* L. at different growth stages ranged from $10^9$ to $10^{13}$ CFU/g, highlighting the potential variability in bacterial populations depending on plant species, environmental factors, and growth stages [29]. Moreover, previous research on the rhizosphere of five different *Acacia* species also revealed that each sample possessed a unique microbial population and community structure [30]. The rhizosphere is the area surrounding plant roots, characterized by high biological activity and a limited spatial extent. Plant roots secrete various compounds that attract microbial colonization in this region, significantly influencing bacterial colonization [31]. The carbon fixed during plant photosynthesis is partially transported to the roots and released as exudates, creating a nutrient-rich environment that fosters bacterial growth [26,31].

Similarly, endophytic bacterial populations in *Curcuma* stems and leaves also exhibited differences depending on growth stage, cultivation conditions, and plant species or strain. Endophytic bacteria may influence *Curcuma* physiology in several ways. Some strains can enhance curcumin production by inducing plant defense pathways and secondary metabolite biosynthesis [32]. Additionally, endophytes that produce phytohormones can stimulate root development and branching, thereby improving nutrient acquisition [33]. Although the overall bacterial densities in stems ($10^5$ CFU/g) were approximately ten times higher than those in leaves ($10^4$ CFU/g), this difference is relatively moderate. It may be attributed to anatomical and environmental factors; stems are generally more enclosed and structurally supportive, potentially offering a more stable microenvironment for bacterial colonization. In contrast, leaves are more exposed to sunlight, temperature fluctuations, and desiccation, which may limit bacterial persistence [34]. This may be due to differences in structural and chemical barriers, as well as the presence of specific compounds that support bacterial proliferation in stems but not in leaves [35]. Moreover, the higher bacterial populations in plastic house conditions compared to field conditions, especially in *C. amada*, indicate that environmental stability and reduced exposure to external stressors enhance bacterial colonization within plant tissues. The middle growth stage consistently showed the highest endophytic bacterial populations across both stems and leaves, mirroring trends observed in the rhizosphere. However, the rhizosphere bacteria populations could also differ depending on their growth stages and the specific *Curcuma* species or strains [36]. This pattern suggests that the internal plant environment at this stage is particularly conducive to bacterial growth [37]. Physiological changes in the plant, such as increased vascular activity and greater nutrient transport, likely facilitate bacterial movement and proliferation [38]. The early stage showed relatively low bacterial numbers, likely due to underdeveloped vascular structures limiting microbial infiltration, while the late stage exhibited a slight decline, possibly due to structural lignification or reduced nutrient flow [39]. Additionally, as plants grow, their immune responses fluctuate, and at the middle stage, they may provide a transiently permissive environment for bacterial colonization before later adopting more robust defense mechanisms [39,40].

A key observation from this study is the significantly higher bacterial population in the stems and leaves of *C. amada* and Ryudai Gold turmeric compared to other *Curcuma* species or strains. This suggests that these strains possess specific traits, such as enhanced production of bacterial-attracting metabolites or structural features that facilitate bacterial entry and colonization. Previous studies have demonstrated that different plant species, and even cultivars within the same species, harbor distinct microbial communities due to variations in metabolic composition [41,42].

The differences in endophytic bacterial populations between stems and leaves further highlight the influence of plant physiology on microbial colonization. Stems, being directly connected to roots through the vascular system, likely serve as the primary conduit for bacterial migration from the rhizosphere [43]. Once inside the plant, bacteria may move through the xylem, assisted by transpiration flow and bacterial motility mechanisms such as flagella [23]. Conversely, the lower bacterial population in leaves may be attributed to their relatively harsher environment, with exposure to UV radiation, fluctuating temperatures, and stronger plant defense responses [44]. The possibility of bacterial colonization through stomatal entry or wounds suggests an additional route for endophytes in leaves, but this appears to be less efficient than root-mediated colonization [45].

In addition to environmental factors, the morphological and phenotypic differences among the *Curcuma* species and strains studied may contribute to the observed variations in microbial populations [46]. For example, strains such as *C. amada* and *C. longa* exhibited larger rhizomes and thicker stems compared to other species, which could provide more extensive surface area and a richer microenvironment for microbial colonization [47]. These differences in plant structure may create unique niches that favor specific bacterial communities in the rhizosphere and within the plant tissues. Furthermore, distinct leaf morphology in species such as *C. zedoaria* and *C. xanthorrhiza* could influence the types of microbial communities present in the leaves [48]. The more compact or waxy surface of the leaves may alter microbial attachment and nutrient availability, potentially resulting in a different microbiome compared to species with thinner or more porous leaves [49].

The PCA analysis further reinforced the influence of growth stage on bacterial population distribution in *Curcuma* species cultivated in field and plastic house, as reflected in the distinct separations observed in the score and loading plots. In the field, *C. aromatica* clustered distinctly in the negative direction of both PC1 and PC2, suggesting highly unique bacterial populations, likely driven by specific root exudates and secondary metabolites that vary across growth stages. Similarly, *C. longa* strain L2 separated from strains L1 and L3, indicating that even within a species, bacterial population differences may arise due to genetic traits and growth stage influences [39]. The loading plot further underscored the impact of anatomical parts on bacterial diversity. Rhizosphere bacterial populations were strongly associated with high positive loadings along PC1, emphasizing their dominance and distinct contribution to microbial composition. In contrast, bacterial populations from leaves and stems showed greater overlap, indicating less distinct microbial communities in these anatomical parts compared to the rhizosphere [19].

In the plastic house, the interaction between growth stage and anatomical parts was less pronounced due to the controlled conditions reducing environmental variability. *C. longa* strains L1, L2, L3, RG, and *C. aromatica* formed a close cluster along PC1 and PC2, indicating a convergence of bacterial population trends that could be influenced by consistent soil, temperature, and moisture conditions [50]. Despite this homogenization, the rhizosphere continued to serve as the microbial hotspot, with strong positive loadings along PC1 across all species and growth stages. PCA results showed that *C. amada* rhizosphere samples contributed strongly to the separation along the positive PC1 and PC2 axes, indicating distinct patterns in bacterial community structure. In contrast, bacterial populations in leaves and stems contributed less to the variation, possibly reflecting the reduced microbial differentiation in these parts under controlled conditions [51]. Interestingly, *C. amada* and *C. zedoaria* demonstrated higher bacterial populations in their stem and leaf samples compared to other species, reflecting potential growth stage-specific shifts in microbial recruitment.

Overall, the microbial populations across different growth stages in different *Curcuma* species exhibit distinct variations between the field experiment and plastic house experiment. The distinct clustering patterns observed in both experiments highlight the critical role of the growth stage in shaping bacterial populations across anatomical parts. In the field, the dynamic shifts in bacterial diversity at different growth stages were dominated by early-stage opportunists, mid-stage metabolic activity, and late-stage specialization, as evidenced by the broader distribution of *Curcuma* species along PC1 and PC2. This variability reflects the influence of heterogeneous environmental factors and stage-specific plant physiological changes on microbial recruitment [52]. In contrast, the plastic house experiment reduced this variability, as seen in the tighter clustering among *Curcuma* species. However, rhizosphere bacterial populations remained a key driver of separation, while leaf and stem populations showed greater overlap across growth stages and species. This pattern highlights that while the rhizosphere remains highly dynamic, the controlled conditions homogenize bacterial communities associated with aerial parts. The strong alignment between PCA clustering and bacterial count data provides robust support for the hypothesis that differed growth stages could influence the bacterial population across all plant compartments and environmental conditions, synergistically shaping microbial diversity patterns.

While this study provides valuable insights into the microbial population dynamics in *Curcuma* species across different growth stages and planting environments, the limitations should be acknowledged, including the absence of microbial community profiling using molecular techniques, and the lack of endophytic fungal identification, which may limit a deeper understanding of microbial diversity and functional interactions within the plant. This study demonstrates that even though different *Curcuma* species were grown under identical conditions (field and plastic house), the microbial populations can still differ significantly, highlighting the complex interactions between the species and their respective environments. The overall findings of this study describe how growth stage and cultivation environment play a crucial role in shaping bacterial populations associated with *Curcuma* species. The peak bacterial abundance observed in the middle stage across all compartments highlights the significance of plant metabolic activity in microbial recruitment. The differences in microbial composition between field and plastic house conditions further demonstrate the impact of external environmental factors on bacterial colonization patterns. These insights contribute to our understanding of plant–microbe interactions in medicinal plants and suggest potential strategies for optimizing *Curcuma* cultivation to enhance beneficial microbial associations.

## 5. Conclusions

This study evaluated the endophytic and rhizosphere bacterial populations across different growth stages in various *Curcuma* species and strains cultivated under field and plastic house conditions. The results revealed notable differences in bacterial populations, with higher densities of endophytic bacteria observed in the leaves and stems during the middle growth stages of all *Curcuma* species or strains. Rhizosphere bacterial populations also varied depending on growth stage and plant species or strain. Understanding these dynamics provides valuable insights for optimizing microbial inoculant application strategies and developing sustainable agricultural practices tailored to specific crop stages and species. Future studies should explore the functional roles of these microbial communities and their potential applications as biofertilizer and stimulant in promoting plant health and productivity. Moreover, incorporating molecular identification and community profiling techniques are needed to uncover the taxonomic and functional diversity of these microbial communities.

**Author Contributions:** Experiment management, data collection, data analysis and manuscript writing, N.I.R.; experiment planning and manuscript editing, M.A.H.; manuscript review and editing, H.A. All authors have read and agreed to the published version of the manuscript.

**Funding:** This research received no external funding.

**Institutional Review Board Statement:** Not applicable.

**Data Availability Statement:** The original contributions presented in this study are included in the article. Further inquiries can be directed to the corresponding author.

**Acknowledgments:** All the authors have approved that this manuscript is applicable for the doctoral degree of the first author, Neptu Islamy Raharja, and express their gratitude to MEXT, Japan, for providing a scholarship to the first author.

**Conflicts of Interest:** The authors declare that they have no conflicts of interest.

## References

1. Kaliyadasa, E.; Samarasinghe, B.A. A review on golden species of *Zingiberaceae* family around the world: Genus *Curcuma*. *Afr. J. Agric. Res.* **2019**, *14*, 519–531.
2. Jyotirmayee, B.; Mahalik, G. A review on selected pharmacological activities of *Curcuma longa* L. *Int. J. Food Prop.* **2022**, *25*, 1377–1398. [CrossRef]
3. de Oliveira Filho, J.G.; de Almeida, M.J.; Sousa, T.L.; dos Santos, D.C.; Egea, M.B. Bioactive compounds of turmeric (*Curcuma longa* L.). In *Bioactive Compounds of Under-Utilized Vegetables and Legumes*; Alexandru, M.G., Holban, A.M., Eds.; Academic Press: Cambridge, MA, USA, 2021; pp. 297–318.
4. Gaiero, J.R.; McCall, C.A.; Thompson, K.A.; Day, N.J.; Best, A.S.; Dunfield, K.E. Inside the root microbiome: Bacterial root endophytes and plant growth promotion. *Am. J. Bot.* **2013**, *100*, 1738–1750. [CrossRef] [PubMed]
5. Bacon, C.W.; White, J.F. Functions, mechanisms and regulation of endophytic and epiphytic microbial communities of plants. *Symbiosis* **2016**, *68*, 87–98. [CrossRef]
6. Khan, S.; Ambika Rani, K.; Sharma, S.; Kumar, A.; Singh, S.; Singh, Y. Rhizobacterial-mediated interactions in *Curcuma longa* for plant growth and enhanced crop productivity: A systematic review. *Front. Plant Sci.* **2023**, *14*, 1231676. [CrossRef]
7. Schrey, D.; Hartmann, A.; Hampp, R. Rhizosphere interactions. In *Ecological Biochemistry: Environmental and Interspecies Interactions*, 2nd ed.; Wink, M., Ed.; Wiley-Blackwell: Weinheim, Germany, 2014; pp. 292–311.
8. Song, Q.; Deng, X.; Song, R.; Song, X. Plant growth-promoting rhizobacteria promote growth of seedlings, regulate soil microbial community, and alleviate damping-off disease caused by *Rhizoctonia solani* on *Pinus sylvestris* var. *mongolica*. *Plant Dis.* **2022**, *106*, 2730–2740. [CrossRef]
9. Vocciante, M.; Grifoni, M.; Fusini, D.; Petruzzelli, G.; Franchi, E. The role of plant growth-promoting rhizobacteria (PGPR) in mitigating plant environmental stresses. *Appl. Sci.* **2022**, *12*, 1231. [CrossRef]
10. Nadeem, S.M.; Ahmad, M.; Zahir, Z.A.; Javaid, A.; Ashraf, M. The role of mycorrhizae and plant growth-promoting rhizobacteria (PGPR) in improving crop productivity under stressful environments. *Biotechnol. Adv.* **2014**, *32*, 429–448. [CrossRef]
11. Compant, S.; Cambon, M.C.; Vacher, C.; Mitter, B.; Samad, A.; Sessitsch, A. The plant endosphere world—Bacterial life within plants. *Environ. Microbiol.* **2021**, *23*, 1812–1829. [CrossRef]
12. Sontsa-Donhoung, A.M.; Bahdjolbe, M.; Hawaou; Nwaga, D. Selecting endophytes for rhizome production, curcumin content, biocontrol potential, and antioxidant activities of turmeric (*Curcuma longa*). *Biomed. Res. Int.* **2022**, *2022*, 8321734. [CrossRef]
13. Thapa, S.; Prasanna, R. Prospecting the characteristics and significance of the phyllosphere microbiome. *Ann. Microbiol.* **2018**, *68*, 229–245. [CrossRef]
14. Manikandan, A.; Anandham, R.; Madhan, S.; Raghu, R.; Krishnamoorthy, R.; Senthilkumar, M. Exploring the phyllosphere: Microbial diversity, interactions, and ecological significance in plant health. In *Plant–Microbe Interactions in Stress Management*; Giri, B., Varma, A., Eds.; Springer: Singapore, 2024; pp. 29–49.
15. Sumathi, C.S.; Balasubramanian, V.; Ramesh, N.; Kannan, V.R. Influence of biotic and abiotic features on *Curcuma longa* L. plantation under tropical condition. *Middle-East J. Sci. Res.* **2008**, *3*, 171–178.
16. Harish, B.S.; Umesha, K.; Venugopalan, R.; Prasad, B.M. Photo-selective nets influence physiology, growth, yield and quality of turmeric (*Curcuma longa* L.). *Ind. Crops Prod.* **2022**, *186*, 115202. [CrossRef]
17. Ishimine, Y.; Hossain, M.; Murayama, S. Optimal planting depth for turmeric (*Curcuma longa* L.) cultivation in dark red soil in Okinawa Island, southern Japan. *Plant Prod. Sci.* **2003**, *6*, 83–89. [CrossRef]
18. Hossain, M.A. Effect of harvest time on shoot biomass and yield of turmeric (*Curcuma longa* L.) in Okinawa, Japan. *Plant Prod. Sci.* **2010**, *13*, 97–103. [CrossRef]
19. Kumar, A.; Singh, R.; Yadav, A.; Giri, D.D.; Singh, P.K.; Pandey, K.D. Isolation and characterization of bacterial endophytes of *Curcuma longa* L. *3 Biotech* **2016**, *6*, 60. [CrossRef]
20. Song, Y.; Li, X.; Yao, S.; Yang, X.; Jiang, X. Correlations between soil metabolomics and bacterial community structures in the pepper rhizosphere under plastic greenhouse cultivation. *Sci. Total Environ.* **2020**, *728*, 138439. [CrossRef]
21. Vives-Peris, V.; de Ollas, C.; Gomez-Cadenas, A.; Perez-Clemente, R.M. Root exudates: From plant to rhizosphere and beyond. *Plant Cell Rep.* **2020**, *39*, 3–17. [CrossRef]
22. Chaparro, J.M.; Badri, D.V.; Vivanco, J.M. Rhizosphere microbiome assemblage is affected by plant development. *ISME J.* **2014**, *8*, 790–803. [CrossRef]

23. Compant, S.; Clément, C.; Sessitsch, A. Plant growth-promoting bacteria in the rhizo- and endosphere of plants: Their role colonization, mechanisms involved and prospects for utilization. *Soil Biol. Biochem.* **2010**, *42*, 669–678. [CrossRef]
24. Reinhold-Hurek, B.; Bünger, W.; Burbano, C.S.; Sabale, M.; Hurek, T. Roots shaping their microbiome: Global hotspots for microbial activity. *Annu. Rev. Phytopathol.* **2015**, *53*, 403–424. [CrossRef] [PubMed]
25. Compant, S.; Duffy, B.; Nowak, J.; Clément, C.; Barka, E.A. Use of plant growth-promoting bacteria for biocontrol of plant diseases: Principles, mechanisms of action, and future prospects. *Appl. Environ. Microbiol.* **2005**, *71*, 4951–4959. [CrossRef] [PubMed]
26. Lugtenberg, B.; Kamilova, F. Plant-growth-promoting rhizobacteria. *Annu. Rev. Microbiol.* **2009**, *63*, 541–556. [CrossRef] [PubMed]
27. Bulgarelli, D.; Schlaeppi, K.; Spaepen, S.; Ver Loren van Themaat, E.; Schulze-Lefert, P. Structure and functions of the bacterial microbiota of plants. *Annu. Rev. Plant Biol.* **2013**, *64*, 807–838. [CrossRef]
28. Kaskoos, R.A.; Ahamad, J.; Mir, S.R.; Javed, S.A. Antimicrobial properties of curcumin and its potential in the treatment of infections. *Pharm. Biol.* **2013**, *51*, 607–611.
29. Cavaglieri, L.; Orlando, J.; Etcheverry, M. Rhizosphere microbial community structure at different maize plant growth stages and root locations. *Microbiol. Res.* **2009**, *164*, 391–399. [CrossRef]
30. Yuan, Z.S.; Liu, F.; He, S.B.; Zhou, L.I.; Pan, H. Community structure and diversity characteristics of rhizosphere and root endophytic bacterial community in different *Acacia* species. *PLoS ONE* **2022**, *17*, e0262909. [CrossRef]
31. Upadhyay, S.K.; Srivastava, A.K.; Rajput, V.D.; Chauhan, P.K.; Bhojiya, A.A.; Jain, D.; Chaubey, G.; Dwivedi, P.; Sharma, B.; Minkina, T. Root exudates: Mechanistic insight of plant growth-promoting rhizobacteria for sustainable crop production. *Front. Microbiol.* **2022**, *13*, 916488. [CrossRef]
32. Singh, L.P.; Gill, S.S.; Tuteja, N. Unraveling the role of fungal endophytes in plant growth and stress tolerance. *Plant Signal. Behav.* **2020**, *15*, 1780043.
33. Santoyo, G.; Moreno-Hagelsieb, G.; del Carmen Orozco-Mosqueda, M.; Glick, B.R. Plant growth-promoting bacterial endophytes. *Microbiol. Res.* **2016**, *183*, 92–99. [CrossRef]
34. Kandel, S.L.; Joubert, P.M.; Doty, S.L. Bacterial endophyte colonization and distribution within plants. *Microorganisms* **2017**, *5*, 77. [CrossRef] [PubMed]
35. Meng, F.C.; Zhou, Y.Q.; Ren, D.; Wang, R.; Wang, C.; Lin, L.G.; Zhang, X.Q.; Ye, W.C.; Zhang, Q.W. Turmeric: A review of its chemical composition, quality control, bioactivity, and pharmaceutical application. In *Handbook of Food Bioengineering*; Volume 7, Holban, A.M., Grumezescu, A.M., Eds.; Academic Press: Cambridge, MA, USA, 2018; pp. 299–350.
36. Philippot, L.; Raaijmakers, J.M.; Lemanceau, P.; van der Putten, W.H. Going back to the roots: The microbial ecology of the rhizosphere. *Nat. Rev. Microbiol.* **2013**, *11*, 789–799. [CrossRef] [PubMed]
37. Marag, P.S.; Suman, A. Growth stage and tissue-specific colonization of endophytic bacteria having plant growth-promoting traits in hybrid and composite maize (*Zea mays* L.). *Microbiol. Res.* **2018**, *214*, 101–113. [CrossRef] [PubMed]
38. Compant, S.; Reiter, B.; Sessitsch, A.; Nowak, J.; Clément, C.; Barka, E.A. Endophytic colonization of *Vitis vinifera* L. by plant growth-promoting bacterium *Burkholderia* sp. strain PsJN. *Appl. Environ. Microbiol.* **2005**, *71*, 1685–1693. [CrossRef]
39. Awin, T.; Mediani, A.; Leong, S.W.; Faudzi, S.M.M.; Shaari, K.; Abas, F. Phytochemical and bioactivity alterations of *Curcuma* species harvested at different growth stages by NMR-based metabolomics. *J. Food Compos. Anal.* **2019**, *77*, 66–76. [CrossRef]
40. Mengistu, A.A. Endophytes: Colonization, behaviour, and their role in defense mechanisms. *Int. J. Microbiol.* **2020**, *2020*, 6927219. [CrossRef]
41. Graner, G.; Paula, P.; Johan, M.; Sadhna, A. Microbial diversity in different cultivars of *Brassica napus* in relation to its wilt pathogen *Verticillium longisporum*. *FEMS Microbiol. Lett.* **2003**, *224*, 269–276. [CrossRef]
42. Monika, S.; Prem, P.S.; Arun, K.P.; Singh, P.K.; Pandey, K.D. Enumeration of culturable endophytic bacterial population of different *Lycopersicon esculentum* L. varieties. *Int. J. Curr. Microbiol. Appl. Sci.* **2018**, *7*, 3344–3352.
43. Wheatley, R.M.; Poole, P.S. Mechanisms of bacterial attachment to roots. *FEMS Microbiol. Rev.* **2018**, *42*, 448–461. [CrossRef]
44. Beattie, G.A.; Lindow, S.E. The secret life of foliar bacterial pathogens on leaves. *Annu. Rev. Phytopathol.* **1995**, *33*, 145–172. [CrossRef]
45. Jin, H.; Yang, X.Y.; Yan, Z.Q.; Liu, Q.; Li, X.Z.; Chen, J.X.; Qin, B. Characterization of rhizosphere and endophytic bacterial communities from leaves, stems, and roots of medicinal *Stellera chamaejasme* L. *Syst. Appl. Microbiol.* **2014**, *37*, 376–385. [CrossRef]
46. Compant, S.; Samad, A.; Faist, H.; Sessitsch, A. A review on the plant microbiome: Ecology, functions, and emerging trends in microbial application. *J. Adv. Res.* **2019**, *19*, 29–46. [CrossRef] [PubMed]
47. Reinhold-Hurek, B.; Hurek, T. Living inside plants: Bacterial endophytes. *Curr. Opin. Plant Biol.* **2011**, *14*, 435–443. [CrossRef]
48. Hardoim, P.R.; van Overbeek, L.S.; van Elsas, J.D. Properties of bacterial endophytes and their proposed role in plant growth. *Trends Microbiol.* **2008**, *16*, 463–471. [CrossRef] [PubMed]
49. Hunter, P.J.; Hand, P.; Pink, D.; Whipps, J.M.; Bending, G.D. Both leaf properties and microbe–microbe interactions influence within-species variation in bacterial population diversity and structure in the lettuce phyllosphere. *Appl. Environ. Microbiol.* **2010**, *76*, 8117–8125. [CrossRef] [PubMed]

50. Kumar, A.; Singh, R.; Yadav, A.; Giri, D.D.; Singh, P.K.; Pandey, K.D. Diversity of bacterial endophytes from *Curcuma longa* L. and their role in plant growth promotion and antifungal activity. *J. Appl. Microbiol.* **2016**, *120*, 1653–1665.
51. Aswathy, R.G.; Joshi, S.R.; Jha, D.K. *Paenibacillus* sp.: A novel endophyte from *Curcuma longa* with potential plant growth-promoting properties. *Res. J. Microbiol.* **2013**, *8*, 115–123.
52. Gumiere, T.; Durrer, A.; Bohnen, H.; Vargas, L.K. Plant growth stage drives temporal and spatial dynamics of the bacterial microbiome in the rhizosphere of *Vigna subterranea*. *Front. Microbiol.* **2022**, *13*, 825377.

*agriculture*

MDPI

*Article*

# Arbuscular Mycorrhizal Fungi Inoculation and Water Regime Effects on Seedling P Uptake by Rice and Pearl Millet

**Phoura Y and Akihiko Kamoshita ***

Asian Research Center for Bioresource and Environmental Sciences (ARC-BRES), Graduate School of Agricultural and Life Sciences, The University of Tokyo, 1-1-1 Midoricho Nishitokyo, Tokyo 188-0002, Japan; phouray.rupp@gmail.com

* Correspondence: akamoshita@g.ecc.u-tokyo.ac.jp; Tel.: +81-70-1579-9740

**Abstract:** Mycorrhizal-mediated seedling establishment may reduce dependency on chemical fertilizers, but the effectiveness of infection for growth may differ depending on species with different eco-physiological adaptations. The infection of arbuscular mycorrhizal fungi (AMF) and P uptake were compared between rice (*Oryza sativa* L.) (Koshihikari (rice$_k$), Togo4 (rice$_t$)), and pearl millet (*Pennisetum glaucum* (L.) R. Br.) (ICMB89111 (millet$_{891}$), ICMB95444 (millet$_{954}$)) seedlings (i) in response to three different commercial AMF inoculants of *Rhizoglomus irregulare* (popular inoculant Dr. Kinkon ($I_1$); two new inoculants Rootella P ($I_2$) and Rootella F ($I_3$)) in comparison with indigenous AMF from Andosol upland and paddy topsoils (Exp. 1–2 as the inoculant experiments) and (ii) across different water regimes from upland to flooded lowland conditions for $I_1$ inoculant (Exp. 3–4 as the water regime experiments). The new inoculants $I_2$ and $I_3$ with higher propagule numbers showed a higher infection rate than the control seedlings in both rice and pearl millet, with a tendency for slower leaf development and no seedling growth enhancement. $I_1$ inoculant had more significant positive effects on the root transversal area and shoot growth parameters than the control. The infection rates of all three inoculants were lower than the indigenous AMF from upland Andosol in rice and pearl millet, in which a higher infection rate led to higher P uptake found in millet$_{954}$. $I_1$ inoculant increased the infection rate in pearl millet and rice but had no clear indication of interaction with water regimes. A higher infection rate led to higher P uptake and shoot dry weight in pearl millet but not in rice with higher root length density. This study provided the significance of inoculants for seedling establishment and highlighted more mycorrhizal-mediated P uptake in pearl millet than in rice.

**Keywords:** arbuscular mycorrhizal fungi; inoculation; pearl millet (*Pennisetum glaucum* (L.) R. Br.); rice (*Oryza sativa* L.); water regime

## 1. Introduction

The mycorrhizal-mediated seedling establishment may reduce dependency on chemical fertilizers, but the effectiveness of inoculants and their interaction with water regimes are poorly understood. Arbuscular mycorrhizal fungi (AMF) may help plant establishment by providing nutrients and water supply to the host crops [1]. Infection rates may differ between commercial inoculants and soil-indigenous AMF, and there may be an optimal level of water regimes for effective inoculation.

AMF colonization is known to start from the first week after inoculation; germinated spores could develop external hyphae in 1 day, internal hyphae after the entry point in 2 days, and then arbuscules and vesicles in ~4 days [2]. Inoculation with AMF in a seeding bed improved plant growth in the seedling stage (3–4 weeks old as in rice and pearl millet seedlings (knowledgebank.irri.org, accessed on 19 March 2025; [3])), which led to superior growth after transplanting in the fields, higher nutrient uptake, and higher yield [4]. The infection rate of 4-week-old rice seedlings reached 40% at maximum [4], which is usually much lower and varies depending on the various factors (e.g., water regimes, types of inoculants, the number of inoculant spores or propagules, and P availability). For example, an upland nursery (a 60% water holding capacity) had a three times higher infection rate than a flooded nursery (a 3 cm water depth) [4]. Auge reviewed inoculation effects under drought in many plants, showing positive responses such as growth, plant water status, and stomata conductance in some cases [5]. However, neither rice nor pearl millet was covered.

AMF commonly resides in the soil as indigenous. Commercial AMF inoculants with some AMF species are available. Many commercial inoculants have been used and reported internationally; for example, Salomon et al. [6] reported nearly 30 inoculants. *Rhizoglomus irregulare*, formerly known as *Rhizophagus irregularis* or *Glomus intraradices* [7], is the model AMF due to its ability to produce high P uptake [8] and propagation in vitro. In Japan, only a few products have been used, among which Dr. Kinkon (*Glomus* sp. strain R-10) (Idemitsu Kosan Co., Ltd., Tokyo, Japan) (https://www.idemitsu.com/jp/content/100038434.pdf, accessed on 19 March 2025) has been the most popular product in Japan for over 20 years; it was tested on wild legume [9], onion [10], and soybean [11]. Recently, sets of new products from Israel were introduced in Japan as Rootella with *R. irregularis* (Groundwork BioAg Co., Ltd., Moshav Mazor, Israel, accessed on 27 March 2025), which claimed to have positive effects on growth, yield, nutrient uptake, and response to stresses to many crop species. Several types of Rootella products are available (e.g., Rootella P (strain P-type), Rootella F (strain F type)), whose propagule numbers are much higher than Dr. Kinkon and may increase infection further. It should not be neglected that natural soils in fields could have large numbers of indigenous AMF that would cause positive plant growth, as reported in Andosol's case by Solaiman and Hirata [12]. It should also be kept in mind that inoculants do not always have positive effects on crop growth [13,14].

Here, this study aimed to make a preliminary investigation of (1) the effects of three different commercial AMF inoculants and indigenous AMF from Andosol upland and paddy soils and (2) the effects of AMF inoculation under different water regimes on the infection and growth of rice and pearl millet at seedling stages. The new inoculant with higher propagule numbers may significantly affect crops and there may be an interaction with water regimes. In addition, rice and pearl millet with different root systems may respond differently to enhanced infection rates.

## 2. Materials and Methods

Three experiments (Exp. 1–3) were simultaneously conducted from June to July 2020, and another experiment (Exp. 4) was conducted from May to June 2021 in the greenhouse at the Institute for Sustainable Agro-ecosystem Services (ISAS), The University of Tokyo, Nishitokyo, Japan (35°43′ N, 139°32′ E). All the experiments were conducted during the seedling stage in cell trays, which were rotated weekly: Exp. 1 with three inoculant types (2020), Exp. 2 with indigenous AMF from two soil types (2020), Exp. 3 with four water regimes with inoculant (2020), and Exp. 4 with five water regimes with inoculant and control (2021).

### 2.1. Plant Materials

Two genotypes (hybrid 'Togo4', 'Koshihikari') of rice (*Oryza sativa* L.) and two (ICMB89111, ICMB95444) of pearl millet (*Pennisetum glaucum* (L.) R. Br.) were used. Koshihikari (rice$_k$) is an improved lowland genotype of *O. sativa* ssp. *japonica* and Togo4 (rice$_t$)

is a high-yielding, good-eating quality hybrid rice with a ricek background, developed by the Research Institute of Rice Production & Technology Co., Ltd., Toyoake, Japan. ICMB89111 (millet$_{891}$) is a high-tillering inbred genotype, and ICMB95444 (millet$_{954}$) is a hybrid genotype, both of which were developed by the International Crop Research Institute for Semi-Arid Tropics (ICRISAT), Hyderabad, India.

### 2.2. Experimental Design

#### 2.2.1. Three Inoculant Types (Exp. 1)

Before sowing, nursery soil containing 29.8% of Akadama (Sharaka Co., Kanuma, Tochigi, Japan), 44.7% of Kanuma (Tomiya Engei, Tenri, Nara, Japan), 17.9% of Ezo Sand (Plantation Iwamoto Co., Hokota, Ibaraki, Japan), and 7.4% of Baido soil (Akagiengei Co., Isesaki, Gunma, Japan) was autoclaved at 121 °C for 30 min to eliminate all the microorganisms, including AMF, which might exist in the nursery soils. Then, the soil was subsampled and oven-dried for three days at 100 °C to determine its gravimetric water content (GWC, %). With no fertilization added, the autoclaved soil was mixed with one of the three commercial inoculants, Dr. Kinkon (R10; $I_1$), Rootella P (strain P-type; $I_2$), or Rootella F (strain F-type; $I_3$), at 10 g/kg of nursery soil based on the products' recommendation. Dr. Kinkon ($I_1$) contained R10 spores, external hyphae, and root fragments with the crystalline-silica carrier and the most probable number of 14 propagules/g with 21 OTUs found in R. irregularis [15]. According to the manufacturer, Rootella P ($I_2$) has at least 2500 viable propagules/g with 88% of clay and 12% of the active ingredient, and Rootella F ($I_3$) has at least 20,800 viable propagules/g with 92% of vermiculite and 8% of the soil amendment ingredient.

Seeds were soaked in cups on 29 June and on 2 July, sowed one per cell (~2 × 2 × 2 cm) in a 100-cell nursery tray (per inoculant treatment, a total of 20 plants for each genotype with 20 additional rice$_k$ as border plants). Each nursery tray with 1.4 kg of soil was placed inside a flat tray, about 1 cm taller than the nursery tray. With an electric scale, the whole trays were weighted and watered daily to maintain a GWC of around 50% based on daily water loss until harvesting on 23 July. Daily minimum and maximum temperatures inside the greenhouse during treatment were 21 and 33 °C.

#### 2.2.2. Indigenous AMF in Two Soil Types (Exp. 2)

Topsoil (0–10 cm) from the paddy lowland (PD) and upland (UP) fields of ISAS in the spring of 2020 was collected, sieved by 0.5 cm mesh to remove debris, and used without autoclaving and adding any fertilizer. The soil chemical properties were as below: the pH ($H_2O$) was 6.6 and 6.6, the total C was 10.6 and 10.4%, the inorganic N was 3.46 and 0.53 mg/100 g, the P was 1.30 and 2.99 mg/100 g, and the K was 21.1 and 49.3 mg/100 g, in PD and UP, respectively. The soils were subsampled and oven-dried for three days at 100 °C to determine their GWC. Growth conditions and dates in Exp. 2 were the same as in Exp. 1.

#### 2.2.3. Water Regimes (Exp. 3 and Exp. 4)

Like Exp. 1, nursery soil in Exp. 3 (29.8% of Akadama, 44.7% of Kanuma, 17.9% of Ezo Sand, and 7.4% of Baido soil) was autoclaved at 121 °C for 30 min and later oven-dried for three days at 100 °C to determine the GWC (%). The autoclaved soil was mixed with $I_1$ for 10 g/kg of nursery soil based on the products' recommendation without adding any fertilization. Until 12 July, growth conditions and dates in Exp. 3 were the same as in Exp. 1. From 13 to 23 July, four water regimes were imposed and monitored daily: water irrigated and maintained at 1 cm above the soil surface (flooded, FL), water held at a similar amount as before treatment (well irrigated, W100), water reduced by 50% of water amount added in W100 (50% well irrigated, W50), and water reduced by 25% of water amount added in

W100 (25% well irrigated, W25). During treatment, whole trays were weighed, and the water amount was added based on the water loss each day. Soil GWC from 13 to 23 July was highest in W100 (53%), W50 (47%), and W25 (41%), respectively. In FL, 30–40% of pearl millet remained for harvest.

In Exp. 4, Kubota nursery soil with N-P-K 0.23–0.32–0.23 g/kg of soil (KN-1U, Kubota Co., Osaka, Japan) was autoclaved at 121 °C for 30 min and oven-dried for three days at 100 °C to determine the GWC (%). The autoclaved soil was added to $I_1$ at 7 g/kg in the inoculation treatment but not in the control treatment. Seeds were soaked on 14 May for pearl millet (again on 16 May due to its poor germination) and 17 May for rice and sowed at one seed per cell (~5 × 5 × 5 cm) in the 50-cell nursery tray (per treatment, a total of 10 plants in each genotype with an additional 5 for $millet_{891}$ and $rice_t$) on 18 May. Each nursery tray was placed inside a flat tray, which was taller by about 1 cm than the nursery tray as in Exp. 3. The whole trays were weighted bidaily with an electric scale, and water amounts were added based on the water loss each day to maintain a GWC of around 30%. To replicate field conditions, from 2 to 16 June, five water regimes were imposed and monitored daily: water irrigated and held at 4 cm above soil surface by placing the tray into a 60 L tank (flooded, FL), water added to saturated (soil saturated, SS), water maintained at the similar amount to that of before treatment (well irrigated, W100), water reduced by 50% of the water amount added in W100 (50% well irrigated, W50), and water reduced by 25% of the water amount added in W100 but bidaily irrigated (25% well irrigated, W25). During the treatment, whole trays were weighed, and the water amount was added based on the water loss each day. Soil GWC from 2 to 16 June was highest in SS (47%), followed by W100 (31%), W50 (28%), and W25 (27%), respectively. Before treatment, a few plants in each genotype were collected for the infection rate, and after treatment, the survival rates of $millet_{891}$ and $millet_{954}$ in FL and SS were 50–60% in the control and 40–50% in the inoculated plants. Daily minimum and maximum temperatures inside the greenhouse during treatment were 17 and 44 °C.

### 2.3. *Measurements*

#### 2.3.1. Shoot Growth

Shoot growth parameters such as tiller number (TN), plant height (PH; cm), leaf age (LA), and shoot dry weight (SDW; g/plant) with 4–5 plants per treatment were collected on 23 July 2020 ($n$ = 4–5) and 16 June 2021 ($n$ = 4). LA was recorded from the first leaf. For SDW, plants were collected and oven-dried for three days at 80 °C.

#### 2.3.2. Shoot Phosphorus Concentration and Uptake

The dry shoots of one or two plants were mixed and ground into powder using a fine mill (Heiko sample mill, TI 300, Fujiwara Seisakusho, Tokyo, Japan) and digested by the MARS 6 microwave (CEM, Matthews, NC, USA) following the modified method note on plant tissue digestion (https://cem.com/ja/digestion-of-plant-tissue-mars-6, accessed on 19 March 2025). The digestion was started by weighting the sample ~0.01–0.1 g by a four-decimal scale before placing it into a digestion vessel, adding 2.5 mL of 68% concentrated nitric acid ($HNO_3$) and 0.5 mL of 30% hydrogen peroxide ($H_2O_2$) in the draft and tightening the lid of the digestion vessels. The microwave was set as follows: Stage: 1; Temp (°C): 220; Ramp (mm: ss): 30:00; Hold (mm: ss): 30:00; Pressure (psi): 800; Power (W): 1200; Stirring: off. After cooling for 20–30 min, the vessel's solution was transferred to a 10 mL tube by rinsing with 2 mL of distilled water twice before filling it up to 5 mL. The solution was mixed and filtrated by a minisart syringe filter (pore size, 0.22 um; Sartorius Co., Goettingen, Germany) up to 4 mL. Standard solutions were prepared as 0, 10, and 50 ppm for phosphorus (P) in three separate 50 mL tubes. The concentration of each standard solution and sample was calibrated and measured by inductively coupled plasma optical emission spectrophotometers (ICP-OES, SPS3000, Hitachi High-Technologies, Tokyo, Japan) following its operation manual. Finally, the concentration (ppm) with dry weight was calculated for nutrient concentration (%) and uptake (mg/plant) ($n$ = 3).

#### 2.3.3. Root Growth

Root length density (RLD; cm/cm$^3$) in 2020 ($n$ = 5) and 2021 ($n$ = 4) was measured from all roots, including the tray bottom. After washing with tap water, roots were immersed in 50% ethanol until measurement. All roots were arranged on a transparent plate and scanned with an Epson Expression 11000XL scanner in professional mode, with positive film, and an 8-bit grayscale at 600 dpi [16]. The images were then analyzed by WinRhizo Pro (Regent Instruments, Quebec, QC, Canada). After scanning, roots at 2–3 cm from the base (one per replication) were cross-sectioned and examined under a phase contrast microscope (BX51, Olympus, Tokyo, Japan). CellSens standard software ver 4.2 (Olympus) was used to capture the microscopic images. Root transversal area (RTA; μm$^2$) was measured by the polygon tool in ImageJ 1.51t (NIH, Bethesda, MD, USA) [17].

#### 2.3.4. Root AMF Infection

A representative sample of roots (0.3–0.8 mm in diameter) in 50% ethanol was stained with the trypan blue method [18]. First, 5–10 pieces of ~2 cm roots were cleared with 10% ($w/v$) KOH by boiling at 110 °C for 15 min, rinsed once with water, soaked in 2% ($w/v$) hydrochloric acid (HCl) at room temperature for ~5 min and dyed with trypan blue solution (0.05% in lactic acid) by boiling at 90 °C for ~5 min. Then, after rinsing thrice with water, roots were stored in lactoglycerol with a ratio of lactic acid/glycerol/water, 8:1:1 ($v/v/v$). Finally, AMF-infected roots were quantified with a modified gridline intersection method [19]. Three roots per replication were placed parallel in a slide glass with a gridline of 1 mm × 1 mm with cover glass and observed at 100× or 200× magnification under a phase-contrast microscope (BX51, Olympus, Hicksville, NY, USA). In total, nine segments of 2 cm roots per 4–5 replications, or 240–300 intersections per treatment, were observed.

At each intersection, the presence or absence of AMF structures was scored for arbuscules, vesicles, and intraradical hyphae. Arbuscules are plant-like structures serving as nutrient exchange sites. Vesicles are oval or circular structures acting as lipid storage rooms, and intraradical hyphae are long, thin, fungal filaments. Arbuscules and vesicles were counted separately, while intraradical hyphae were counted if other mycorrhizal structures were present due to either arbuscules or vesicles implying the presence of hyphae (Figure S1). The mycorrhizal infection rate (or intraradical hyphal infection rate; M%) was calculated by dividing the total number of infected root intersections by the total number of observed root intersections ($n \approx 60$). The arbuscular (A%) and vesicular (V%) infection rates were determined by dividing the number of arbuscule and vesicle intersections by the total number of observed root intersections.

$$\text{M\%} = \frac{Total\ number\ of\ infected\ root\ intersections}{Total\ number\ of\ observed\ root\ intersections} \times 100$$

$$\text{A\%} = \frac{Total\ number\ of\ intersections\ with\ arbuscules}{Total\ number\ of\ observed\ root\ intersections} \times 100$$

$$\text{V\%} = \frac{Total\ number\ of\ intersections\ with\ vesicles}{Total\ number\ of\ observed\ root\ intersections} \times 100$$

### 2.4. Statistical Analysis

Data were analyzed using GenStat 21.1 (VSNi, Hemel, Hempstead, UK). General ANOVA and multiple comparisons (Tukey's test) (significance set at $p < 0.05$) were used to assess the effects of water (4–5 water regimes), inoculation (C, I), genotype (millet$_{891}$, millet$_{954}$, rice$_t$, and rice$_k$), inoculants ($I_1$, $I_2$, and $I_3$), and soils (UP and PD) on infection rates (M%, A%, and V%), P concentration (P%), and uptake and plant growth parameters (SDW, PH, TN, RLD, and RTA). Multiple comparisons for the interactive effects were assessed by the least significant difference (significance set at $p < 0.05$).

## 3. Results

### 3.1. Exp. 1 Inoculation Types

Inoculation affected the infection rate (M%), root growth (RLD and RTA), and shoot growth (SDW, PH, and LA) (Table 1). The M% in $I_2$ was the highest, followed by $I_3$. $I_2$ had the lowest LA and the least SDW. The M% was not significantly different between $I_1$ and control, while $I_1$ had the highest RLD, RTA, SDW, and PH. Rice$_k$ had a higher M% than millet$_{891}$ and rice$_t$, and root and shoot growth parameters were generally larger in rice than pearl millet. The inoculant type interacted with genotypes as $I_1$ enhanced the RLD and SDW for rice$_k$, and $I_2$ decreased the RLD and SDW for rice$_k$ (Table 2).

**Table 1.** Effects of inoculation and genotype on mycorrhizal (M%), arbuscular (A%), and vesicular (V%) infection rates in (%), root length density (RLD, cm/cm$^3$), root transversal area (RTA, $\times 10^4$ μm$^2$), shoot P concentration (P%, %) and P uptake (mg/plant), shoot dry weight (SDW, g/plant), plant height (PH, cm), and leaf age (LA) with control (C) and inoculants Dr. Kinkon ($I_1$) and Rootella P ($I_2$) and F ($I_3$) in pearl millet (ICMB89111, millet$_{891}$; ICMB95444, millet$_{954}$) and rice (Koshihikari, rice$_k$; Togo4, rice$_t$) in 2020 (Exp. 1).

| Parameters | M% | A% | RLD | RTA | P% | P Uptake | SDW | PH | LA |
|---|---|---|---|---|---|---|---|---|---|
| Inoculation (I) | *** | NS | *** | * | NS | NS | *** | NS | *** |
| C | 0.0 ± 0.0 a | 0.0 ± 0.0 | 27 ± 3 ab | 9.6 ± 0.9 ab | 0.146 ± 0.018 | 0.017 ± 0.003 | 0.011 ± 0.000 ab | 11.4 ± 0.2 | 3.8 ± 0.0 b |
| $I_1$ | 2.4 ± 1.1 a | 0.0 ± 0.0 | 31 ± 5 b | 12.5 ± 1.4 b | 0.173 ± 0.027 | 0.023 ± 0.005 | 0.013 ± 0.000 b | 12.9 ± 0.3 | 3.6 ± 0.1 ab |
| $I_2$ | 13.1 ± 3.7 b | 0.0 ± 0.1 | 19 ± 3 a | 10.5 ± 1.3 ab | 0.149 ± 0.009 | 0.014 ± 0.004 | 0.008 ± 0.001 a | 12.2 ± 0.7 | 3.4 ± 0.1 a |
| $I_3$ | 7.5 ± 2.6 ab | 0.1 ± 0.0 | 21 ± 2 a | 8.4 ± 0.8 a | 0.126 ± 0.008 | 0.016 ± 0.003 | 0.012 ± 0.002 b | 12.2 ± 0.8 | 3.5 ± 0.1 a |
| Genotype (G) | * | NS | ** | ** | NS | *** | ** | *** | *** |
| millet$_{891}$ | 2.2 ± 0.9 a | 0.0 ± 0.0 | 17 ± 2 a | 9.1 ± 1.3 a | 0.135 ± 0.018 | 0.010 ± 0.001 a | 0.006 ± 0.002 a | 6.2 ± 1.3 a | 3.3 ± 0.1 a |
| millet$_{954}$ | 5.8 ± 1.9 ab | 0.1 ± 0.1 | 14 ± 3 a | 7.9 ± 0.8 a | 0.140 ± 0.012 | 0.007 ± 0.001 a | 0.004 ± 0.002 a | 5.2 ± 1.7 a | 3.3 ± 0.1 a |
| rice$_k$ | 11.4 ± 4.1 b | 0.0 ± 0.0 | 28 ± 2 b | 10.5 ± 0.9 ab | 0.173 ± 0.027 | 0.024 ± 0.004 b | 0.013 ± 0.002 b | 17.2 ± 1.6 b | 3.9 ± 0.1 b |
| rice$_t$ | 3.6 ± 1.8 a | 0.0 ± 0.0 | 39 ± 4 c | 13.6 ± 1.2 c | 0.147 ± 0.006 | 0.030 ± 0.002 b | 0.016 ± 0.001 c | 17.9 ± 1.5 b | 4.0 ± 0.1 b |
| I×G | NS | NS | * | NS | NS | NS | * | * | NS |

NS, *, **, and *** mean not significant and significant at 5%, 1%, and 0.1% by ANOVA. Different alphabet letters showed significance at 5% by the Tukey multiple comparison test. V% not detected. All parameters were $n = 5$, except for M% and A%, $n = 4$ and P% and P uptake, $n = 3$.

**Table 2.** Inoculation by genotype interaction (IxG) effect on root length density (RLD, cm/cm$^3$) and shoot dry weight (SDW, g/plant) with control (C) and inoculants Dr. Kinkon ($I_1$) and Rootella P ($I_2$) and F ($I_3$) in pearl millet (ICMB89111, millet$_{891}$; ICMB95444, millet$_{954}$) and rice (Koshihikari, rice$_k$; Togo4, rice$_t$) in 2020 (Exp. 1).

| Parameters | RLD | | | | SDW | | | |
|---|---|---|---|---|---|---|---|---|
| | C | $I_1$ | $I_2$ | $I_3$ | C | $I_1$ | $I_2$ | $I_3$ |
| millet891 | 23 ± 4 abc | 21 ± 3 abc | 13 ± 3 ab | 10 ± 2 a | 0.007 ± 0.001 a | 0.007 ± 0.001 a | 0.007 ± 0.000 a | 0.007 ± 0.001 a |
| millet954 | 22 ± 11 abc | 10 ± 6 a | 10 ± 4 a | 14 ± 2 ab | 0.004 ± 0.001 a | 0.003 ± 0.001 a | 0.004 ± 0.001 a | 0.006 ± 0.00 a |
| ricek | 33 ± 5 c | 33 ± 5 c | 22 ± 6 bc | 25 ± 3 bc | 0.016 ± 0.003 bcd | 0.016 ± 0.001 bcd | 0.007 ± 0.002 a | 0.014 ± 0.001 b |
| ricet | 33 ± 5 c | 59 ± 4 d | 33 ± 7 c | 33 ± 4 c | 0.018 ± 0.003 bcd | 0.024 ± 0.002 e | 0.015 ± 0.005 bc | 0.020 ± 0.001 cde |

Different alphabet letters show significance at 5% by least significant difference.

### 3.2. Exp. 2 Soil Types

UP exhibited a higher M% (32.9 vs. 10.2%) and P% (0.417 vs. 0.199%) than PD, while SDW and PH showed higher values in PD with a tendency for greater root growth parameters (RLD, RTA, and not significant) (Table 3). Millet$_{954}$, rice$_k$, and rice$_t$ had a higher M% than millet$_{891}$, whereas shoot growth parameters and RLD were higher in rice than in pearl millet. Rice (rice$_t$ and rice$_k$) exhibited a higher SDW in PD, indicating a significant soil type by genotype interaction (SxG). An increased M% by inoculation led to a higher P uptake for millet$_{954}$ (Figure 1a) but not for rice (Figure 1b). Rice showed higher RLD (Figure 1d) than pearl millet (Figure 1c), and its higher RLD led to higher P uptake. P uptake in pearl millet increased by inoculation to a greater extent in UP and PD than in the other three inoculations and control (Figure 1a).

**Table 3.** Effects of soil type and genotype on mycorrhizal (M%), arbuscular (A%), and vesicular (V%) infection rate (%), root length density (RLD, cm/cm$^3$), root transversal area (RTA, $\times 10^4$ $\mu m^2$), shoot P concentration (P%, %) and P uptake (mg/plant), shoot dry weight (SDW, g/plant), plant height (PH, cm), and leaf age (LA) with indigenous AMF from Andosol upland soil (UP) and paddy soil (PD) in pearl millet (ICMB89111, millet$_{891}$; ICMB95444, millet$_{954}$) and rice (Koshihikari, rice$_k$; Togo4, rice$_t$) in 2020 (Exp. 2).

| Parameters | M% | A% | RLD | RTA | P% | P Uptake | SDW | PH | LA |
|---|---|---|---|---|---|---|---|---|---|
| Soil (S) | ** | NS | NS | NS | ** | NS | *** | *** | *** |
| UP | 32.9 ± 4 b | 2.5 ± 0.4 | 13 ± 4 | 6.9 ± 0.7 | 0.417 ± 0.020 b | 0.015 ± 0.004 | 0.004 ± 0.002 a | 7.4 ± 1.5 a | 2.9 ± 0.1 a |
| PD | 10.2 ± 6 a | 0.4 ± 1.1 | 24 ± 2 | 8.4 ± 0.4 | 0.199 ± 0.085 a | 0.021 ± 0.003 | 0.010 ± 0.000 b | 11.5 ± 1.2 b | 3.7 ± 0.1 b |
| Genotype (G) | *** | NS | NS | NS | NS | * | *** | *** | *** |
| millet$_{891}$ | 4.2 ± 1.8 a | 0.2 ± 0.2 | 12 ± 3 | 7.3 ± 0.9 | 0.249 ± 0.062 | 0.010 ± 0.002 a | 0.005 ± 0.001 a | 4.9 ± 0.8 a | 3.0 ± 0.2 a |
| millet$_{954}$ | 28.9 ± 8.1 b | 2.6 ± 1.5 | 8 ± 2 | 8.2 ± 0.6 | 0.454 ± 0.136 | 0.012 ± 0.003 a | 0.003 ± 0.000 a | 3.5 ± 0.3 a | 3.0 ± 0.1 a |
| rice$_k$ | 28.4 ± 10.9 b | 2.8 ± 1.9 | 27 ± 5 | 5.6 ± 0.6 | 0.22 ± 0.020 | 0.022 ± 0.006 ab | 0.009 ± 0.002 b | 15.1 ± 1.2 b | 3.5 ± 0.2 b |
| rice$_t$ | 24.7 ± 6.6 b | 0.2 ± 0.2 | 27 ± 5 | 9.4 ± 0.9 | 0.31 ± 0.117 | 0.028 ± 0.005 b | 0.011 ± 0.003 b | 14.4 ± 1.6 b | 3.7 ± 0.2 b |
| SxG | NS | NS | NS | NS | NS | NS | * | NS | NS |

NS, *, **, and *** mean not significant and significant at 5%, 1%, and 0.1% by ANOVA. Different alphabet letters show significance at 5% by the Tukey multiple comparison test. V% not detected. All parameters were $n$ = 4–5, except for P% and P uptake, $n$ = 3.

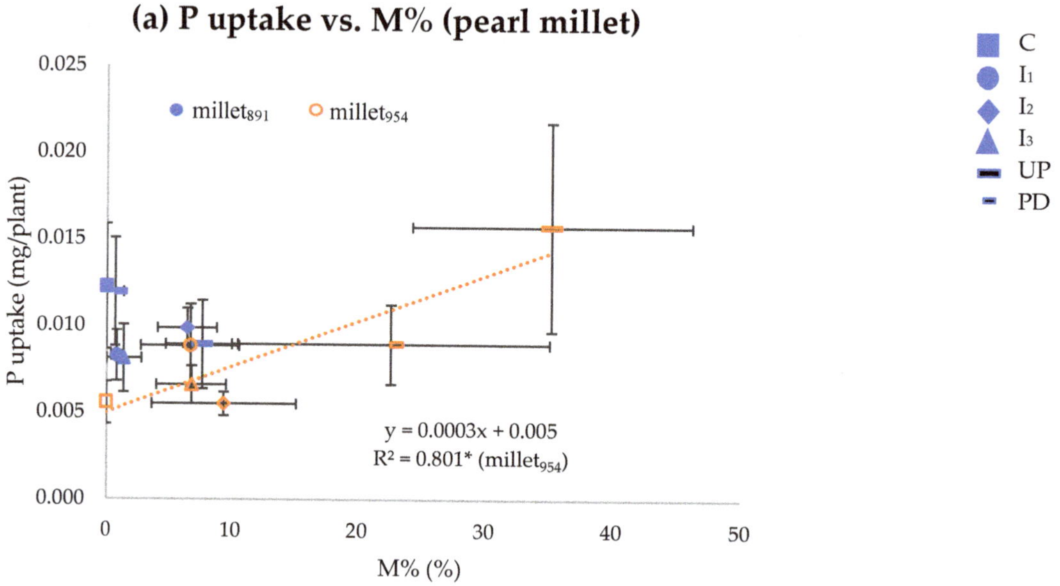


**Figure 1.** *Cont.*

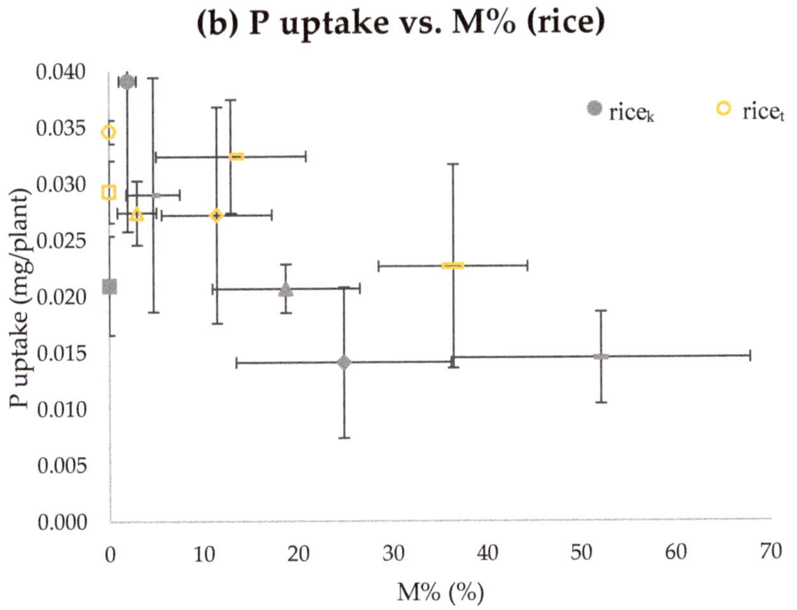


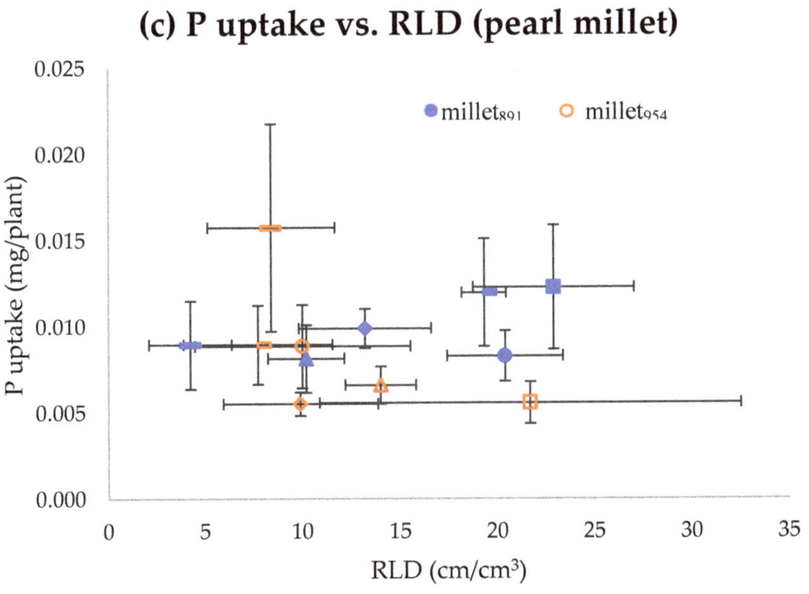


**Figure 1.** *Cont.*

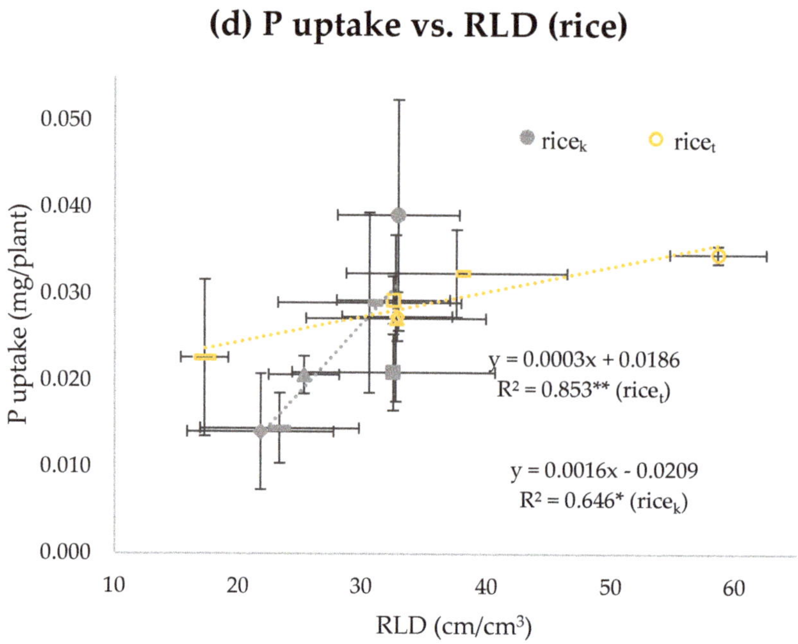


**Figure 1.** Relationships between mycorrhizal infection rate (M%) and P uptake, and root length density (RLD) and P uptake in Exp. 1 and 2 with control (C) and inoculants Dr. Kinkon ($I_1$) and Rootella P ($I_2$) and F ($I_3$), with indigenous AMF from Andosol upland soil (UP) and paddy soil (PD) in pearl millet (ICMB89111, $millet_{891}$; ICMB95444, $millet_{954}$) (**a**,**c**) and rice (Koshihikari, $rice_k$; Togo4, $rice_t$) (**b**,**d**). Error bars indicate standard errors. Linear regressions are drawn when significant at $p < 0.05$ (*) and <0.01 (**).

### *3.3. Exp. 3 and Exp. 4 Water Regimes*

The four water regimes did not affect the infection rates (M% and A%) in Exp. 3, while SDW and PH decreased from FL to W100, W50, and W25; RLD and P uptake declined only under W25 (Table 4). P% was about 0.16% with no effects from the water regimes. Rice had a higher RLD and SDW than pearl millet. The five water regimes also did not affect the infection rates (M%, A%, and V%) in Exp. 4. However, shoot growth parameters (SDW, TN, and PH), P uptake, and RTA were ranked in the order of SS, FL, W100, W50, and W25 from highest to lowest values, while RLD was higher in W50 than FL and W25 (Table 5). P% was higher in W25 and W100 (ca. 0.14%) than SS (0.12%) and W50 (0.11%). Inoculation increased the M%, A%, RLD, RTA, SDW, and TN in Exp. 4. P%, P uptake, and SDW increased in inoculated pearl millet but not in inoculated rice (Table 6). Increasing the M% through inoculation resulted in greater P uptake for pearl millet (Figure 2a) but not for rice (Figure 2b), which exhibited a higher RLD (Figure 2c,d). The M% from inoculation was highest in W50 for pearl millet and W25 for rice, while P uptake from inoculation was highest in SS, followed by W100 in pearl millet and FL and SS in rice (Figure 2a,b).

**Table 4.** Effects of water and genotype on the mycorrhizal (M%), arbuscular (A%), and vesicular (V%) infection rate (%), root length density (RLD, cm/cm$^3$), root transversal area (RTA, $\times 10^4$ μm$^2$), shoot P concentration (P%, %) and P uptake (mg/plant), shoot dry weight (SDW, g/plant), and plant height (PH, cm) with inoculant Dr. Kinkon ($I_1$) in four water availabilities (flooded, FL; well irrigated, W100; 50% well irrigated, W50; 25% well irrigated, W25) in pearl millet (ICMB89111, $millet_{891}$; ICMB95444, $millet_{954}$) and rice (Koshihikari, $rice_k$; Togo4, $rice_t$) in 2020 (Exp. 3).

| Parameters | M% | A% | RLD | RTA | P% | P Uptake | SDW | PH |
|---|---|---|---|---|---|---|---|---|
| Water (W) | NS | NS | *** | *** | NS | ** | *** | *** |
| FL | 1.6 ± 0.9 | 0.0 ± 0.0 | 27 ± 3 b | 12.5 ± 1.2 b | 0.153 ± 0.008 | 0.023 ± 0.004 b | 0.015 ± 0.002 c | 13.9 ± 1.8 b |
| W100 | 2.4 ± 1.1 | 0.0 ± 0.0 | 30 ± 5 b | 12.5 ± 1.4 b | 0.173 ± 0.027 | 0.023 ± 0.005 b | 0.013 ± 0.002 bc | 12.9 ± 1.7 ab |
| W50 | 0.5 ± 1.0 | 0.0 ± 0.6 | 30 ± 2 b | 10.5 ± 0.6 b | 0.166 ± 0.023 | 0.022 ± 0.003 b | 0.011 ± 0.001 ab | 12.6 ± 1.2 ab |
| W25 | 2.5 ± 0.4 | 0.5 ± 0.0 | 19 ± 3 a | 6.2 ± 1.0 a | 0.166 ± 0.011 | 0.011 ± 0.004 a | 0.006 ± 0.001 a | 10.7 ± 1.5 a |
| Genotype (G) | NS | NS | *** | NS | NS | *** | *** | *** |
| $millet_{891}$ | 0.8 ± 0.5 | 0.0 ± 0.0 | 15 ± 2 a | 10.1 ± 1.5 | 0.128 ± 0.016 | 0.008 ± 0.001 a | 0.006 ± 0.001 a | 6.4 ± 0.2 a |
| $millet_{954}$ | 1.7 ± 1.2 | 0.0 ± 0.0 | 17 ± 2 a | 9.1 ± 1.2 | 0.168 ± 0.015 | 0.009 ± 0.001 a | 0.005 ± 0.000 a | 5.7 ± 0.3 a |
| $rice_k$ | 1.9 ± 0.8 | 0.5 ± 0.6 | 35 ± 2 b | 10.3 ± 1.0 | 0.184 ± 0.025 | 0.031 ± 0.004 b | 0.016 ± 0.001 b | 18.3 ± 0.7 b |
| $rice_t$ | 2.6 ± 1.0 | 0.0 ± 0.0 | 40 ± 3 c | 12.2 ± 1.2 | 0.180 ± 0.013 | 0.032 ± 0.002 b | 0.018 ± 0.001 b | 19.7 ± 0.6 c |
| WxG | * | NS | *** | NS | NS | NS | *** | *** |

NS, *, **, and *** mean not significant and significant at 5%, 1%, and 0.1% by ANOVA. Different alphabet letters show significance at 5% by the Tukey multiple comparison test. V% was not detected. All parameters were $n$ = 5, except for M% and A%, $n$ = 4 and P% and P uptake, $n$ = 3.

**Table 5.** Effects of water, inoculation, and genotype on mycorrhizal (M%), arbuscular (A%), and vesicular (V%) infection rate (%), root length density (RLD, cm/cm$^3$), root transversal area (RTA, $\times 10^4$ μm$^2$), shoot P concentration (P%, %) and P uptake (mg/plant), shoot dry weight (SDW, g/plant), plant height (PH, cm), and tiller number (TN) of control (C) and $I_1$ (I) in 5 water regimes (flooded, FL; soil saturated, SS; well irrigated, W100; 50% well irrigated, W50; 25% well irrigated, W25) in pearl millet (ICMB89111, $millet_{891}$; ICMB95444, $millet_{954}$) and rice (Koshihikari, $rice_k$; Togo4, $rice_t$) in 2021 (Exp. 4).

| Parameters | M% | A% | V% | RLD | RTA | P% | P Uptake | SDW | PH | TN |
|---|---|---|---|---|---|---|---|---|---|---|
| Water (W) | NS | NS | NS | *** | *** | ** | *** | *** | *** | *** |
| FL | 0.6 ± 0.2 | 0.0 ± 0.0 | 0.0 ± 0.0 | 9 ± 1 a | 87.4 ± 6.4 d | 0.127 ± 0.007 ab | 0.354 ± 0.062 c | 0.252 ± 0.037 d | 30.5 ± 2.8 d | 2.3 ± 0.2 ab |
| SS | 1.3 ± 0.6 | 0.1 ± 0.1 | 0.1 ± 0.1 | 10 ± 1 ab | 101.8 ± 5.7 e | 0.118 ± 0.008 ab | 0.439 ± 0.072 d | 0.336 ± 0.043 e | 30.8 ± 2.7 d | 2.5 ± 0.3 b |
| W100 | 1.3 ± 0.4 | 0.1 ± 0.1 | 0.0 ± 0.0 | 10 ± 1 ab | 77.0 ± 3.3 c | 0.136 ± 0.01 ab | 0.257 ± 0.028 b | 0.206 ± 0.021 c | 27.5 ± 1.9 c | 2.0 ± 0.2 ab |
| W50 | 2.0 ± 0.4 | 0.1 ± 0.0 | 0.1 ± 0.1 | 11 ± 1 b | 55.8 ± 2.7 b | 0.112 ± 0.01 a | 0.194 ± 0.022 ab | 0.163 ± 0.010 b | 24.8 ± 1.4 b | 1.7 ± 0.1 ab |
| W25 | 1.6 ± 0.6 | 0.0 ± 0.1 | 0.1 ± 0.1 | 9 ± 1 a | 40.5 ± 3.1 a | 0.142 ± 0.005 b | 0.163 ± 0.019 a | 0.108 ± 0.012 a | 19.1 ± 1.4 a | 1.4 ± 0.1 a |
| Inoculation (I) | *** | * | NS | *** | *** | NS | NS | *** | NS | *** |
| C | 0.0 ± 0.0 a | 0.0 ± 0.0 a | 0.0 ± 0.0 | 9 ± 1 a | 67.4 ± 3.8 a | 0.124 ± 0.005 | 0.274 ± 0.035 | 0.198 ± 0.020 a | 26.5 ± 1.4 | 1.9 ± 0.1 a |
| I | 2.7 ± 0.4 b | 0.1 ± 0.0 b | 0.1 ± 0.0 | 10 ± 1 b | 77.6 ± 3.6 b | 0.130 ± 0.005 | 0.289 ± 0.028 | 0.228 ± 0.019 b | 26.6 ± 1.4 | 2.0 ± 0.1 b |
| Genotype (G) | NS | NS | NS | *** | *** | NS | *** | *** | *** | *** |
| $millet_{891}$ | 1.1 ± 0.4 | 0.0 ± 0.0 | 0.0 ± 0.0 | 6 ± 0 a | 59.9 ± 3.9 a | 0.122 ± 0.009 | 0.119 ± 0.013 b | 0.096 ± 0.009 a | 15.4 ± 0.5 a | 2.0 ± 0.0 a |
| $millet_{954}$ | 1.4 ± 0.4 | 0.1 ± 0.1 | 0.0 ± 0.0 | 6 ± 0 a | 76.5 ± 5.4 b | 0.120 ± 0.007 | 0.101 ± 0.013 a | 0.079 ± 0.008 a | 15.6 ± 0.5 a | 2.0 ± 0.0 a |
| $rice_k$ | 1.6 ± 0.5 | 0.1 ± 0.0 | 0.1 ± 0.0 | 14 ± 0 b | 62.5 ± 3.3 a | 0.133 ± 0.009 | 0.440 ± 0.044 c | 0.339 ± 0.025 b | 38.1 ± 1.2 b | 3.0 ± 0.1 b |
| $rice_t$ | 1.3 ± 0.4 | 0.0 ± 0.0 | 0.1 ± 0.0 | 14 ± 0 b | 91.2 ± 6.5 c | 0.134 ± 0.005 | 0.465 ± 0.043 d | 0.339 ± 0.026 b | 37.2 ± 1.2 b | 3.0 ± 0.1 b |
| WxI | NS | NS | NS | NS | NS | *** | NS | NS | *** | NS |
| WxG | NS | NS | NS | * | *** | ** | *** | *** | *** | *** |
| IxG | NS | NS | NS | NS | NS | *** | *** | * | NS | * |
| WxIxG | NS | NS | NS | *** | NS | NS | * | NS | *** | NS |

NS, *, **, and *** mean not significant and significant at 5%, 1%, and 0.1% by ANOVA. Different alphabet letters show significance at 5% by the Tukey multiple comparison test. All parameters were $n$ = 4, except for P% and P uptake, $n$ = 3.

**Table 6.** Effects of IxG on shoot P concentration (P%, %), P uptake (mg/plant), and shoot dry weight (SDW, g/plant) with control (C) and $I_1$ (I) in pearl millet (ICMB89111, $millet_{891}$; ICMB95444, $millet_{954}$) and rice (Koshihikari, $rice_k$; Togo4, $rice_t$) in 2021 (Exp. 4).

| Parameters | P% | | P uptake | | SDW | |
|---|---|---|---|---|---|---|
| | C | I | C | I | C | I |
| $millet_{891}$ | 0.098 ± 0.008 a | 0.145 ± 0.014 c | 0.074 ± 0.014 a | 0.163 ± 0.016 b | 0.071 ± 0.012 a | 0.120 ± 0.011 b |
| $millet_{954}$ | 0.100 ± 0.006 a | 0.140 ± 0.010 b | 0.053 ± 0.009 a | 0.150 ± 0.018 b | 0.047 ± 0007 a | 0.111 ± 0.009 b |
| $rice_k$ | 0.150 ± 0.013 c | 0.117 ± 0.009 ab | 0.488 ± 0.063 d | 0.392 ± 0.061 c | 0.339 ± 0.034 c | 0.339 ± 0.038 c |
| $rice_t$ | 0.145 ± 0.006 c | 0.120 ± 0.005 ab | 0.480 ± 0.062 d | 0.451 ± 0.061 cd | 0.334 ± 0.036 c | 0.343 ± 0.037 c |

Different alphabet letters show significance at 5% by the least square significance test.

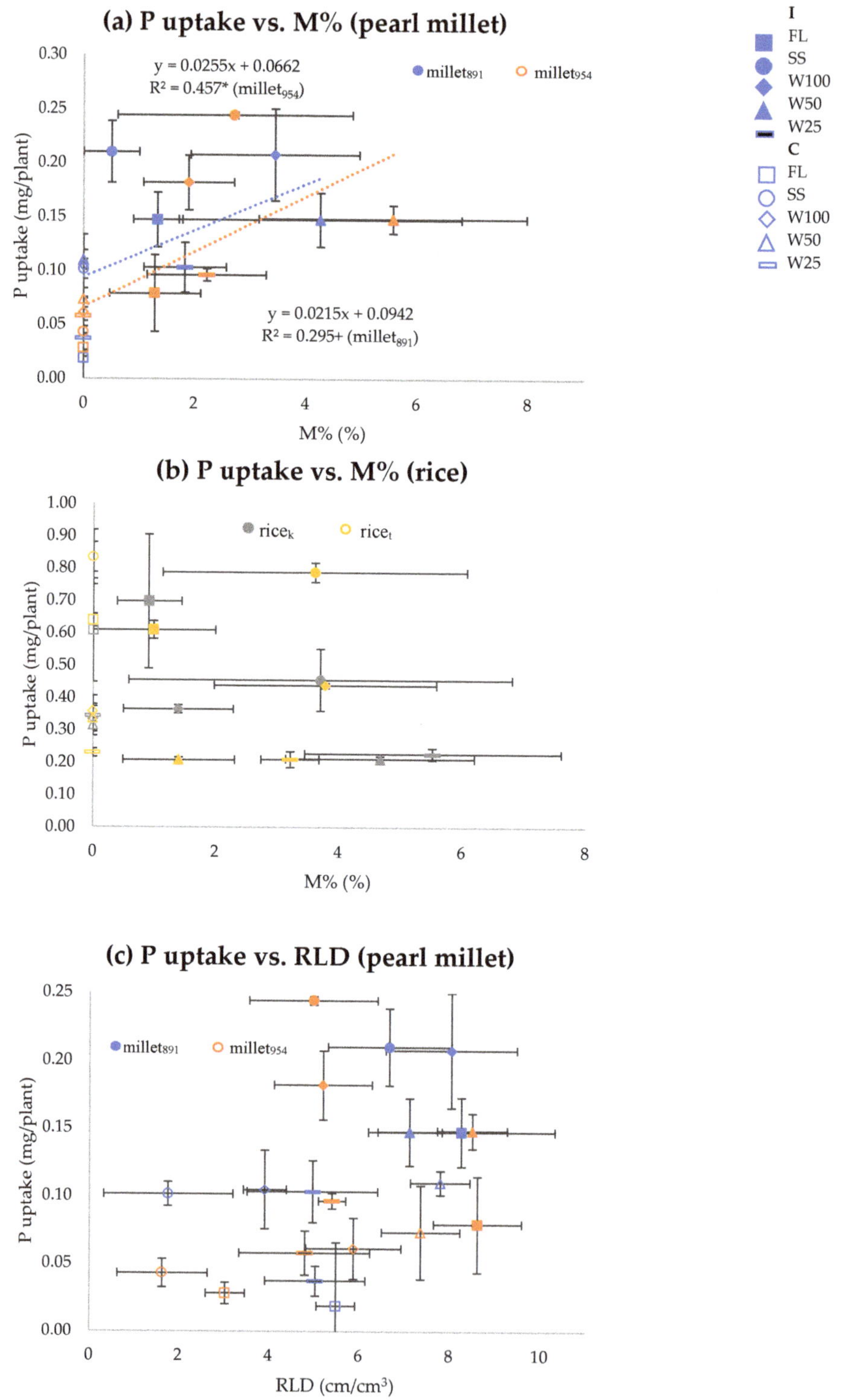


**Figure 2.** *Cont.*

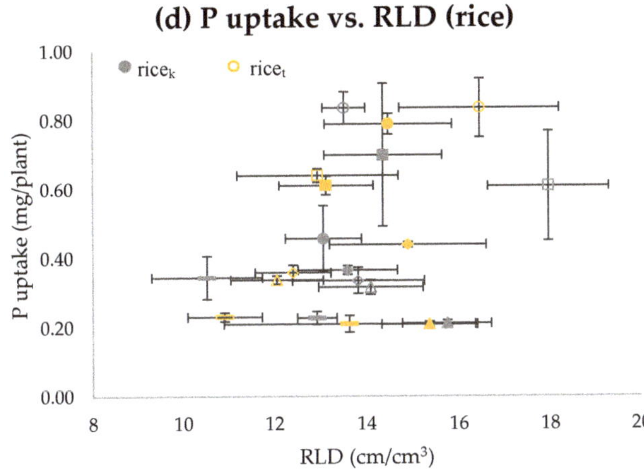


**Figure 2.** Relationships between mycorrhizal infection rate (M%) and P uptake, and root length density (RLD) and P uptake in Exp. 4 for pearl millet (ICMB89111, $millet_{891}$; ICMB95444, $millet_{954}$) (**a**,**c**) and rice (Koshihikari, $rice_k$; Togo4, $rice_t$) (**b**,**d**) with/without inoculation (I, C) across five water regimes (flooded, FL; saturated soil, SS; well irrigated, W100; 50% well irrigated, W50; 25% well irrigated, W25). Error bars indicated standard errors. Linear regressions are drawn when significant at $p < 0.05$ (*).

## 4. Discussion

### *4.1. Inoculation Types and Indigenous AMF*

This study showed small but positive effects of all three inoculants on either infection rates or plant growth parameters with some unique differences between the new inoculants, Rootella ($I_2$ and $I_3$), and the popular Japanese inoculant Dr Kinkon ($I_1$). The common characteristics of all three inoculants included an enhancement in the infection rate of 3-week-old rice and pearl millet seedlings, as well as a slower development (i.e., smaller LA) compared to the control. The new inoculant, Rootella, exhibited a higher M% and smaller LA (Table 1). As part of the assimilates need to be competitively provided to AMF colonization, shoot development might have been delayed in well-infected seedlings [20], as seen from the delayed leaf development. Higher infection rates (i.e., M%) of Rootella would be due to their higher propagule numbers, as indicated by Karasawa et al. [21], but this infection rate was not directly translated into plant growth in the new inoculants. Only the inoculant Dr. Kinkon $I_1$ increased the RTA, SDW, and PH more than the control, whereas Rootella did not show such effects, but $I_2$ showed the suppressed RLD and SDW. The $I_1$ is still uncertain as the carriers attached to $I_1$, like crystalline silica, may or may not have enhanced biomass production, and silica nanoparticles are reported to promote plant growth and improve plant resistance against biotic and abiotic stresses [22]. However, Ortas and Akpinar [23] showed an enhancement of spore production (695 vs. 0 per 100 g soil) and M% (84 vs. 30%) by $I_1$ inoculation over control on six maize genotypes in the pots at the seeding stage at eight weeks, which was comparable or higher than other mycorrhizal inoculations (seven glomus species, one indigenous mycorrhizae, and their cocktail). $I_1$ did not change the shoot dry weight but enhanced the root dry weight and plant concentrations of phosphorus and zinc. Nevertheless, Niwa et al. [15] argued that Dr. Kinkon $I_1$ had typical ruderal traits, which could lead to early infection with mass spore production and good adaptation to agricultural land. Thus, further studies are needed under field conditions to assess the full potential of the new inoculants Rootella with higher propagule density under field conditions since the low infection rates of the current study may have been related to the relatively small volume of the soils in the cell tray. Also, caution is needed when reporting the effects of the commercial mycorrhizal inoculants. From the current inoculation treatment (14 g per tray), the estimated number of propagules could range from 196 ($I_1$) to 350,000 ($I_3$) propagules per tray, according to the product's explanation; however, the actual responding numbers from the internal observation were significantly lower. Salmon et al. [14] also warned that the number of commercial mycorrhizal inoculants did not positively impact root colonization or plant growth.

This study also confirmed that natural soils can become a substantial source of mycorrhizal infection, as was shown by a much higher infection rate from the indigenous AMF collected from the topsoil of Andosol than the autoclaved soils added with the inoculants (Tables 1 and 3; Figure 1). The higher infection rate led to higher P uptake found in $millet_{954}$. It is well known that lowland rice can be infected with AMF but with much lower infection rates than upland rice [24,25], but it was surprising that the natural soils collected from lowland fields also showed substantial infection rates in this experiment. This may be because the soils in the lowland field in this experiment had been fallowed under aerobic conditions from autumn to spring of the previous year when substantial amounts of weeds were infested. However, the soils from the upland field exhibited significantly higher spore and propagule counts of indigenous AMF, as AMF infection measured by qPCR in the upland Andosol was 50% greater than in the lowland Andosol at the same experimental site (K Ejiri, personal communication). From this observation, using indigenous AMF may be one good inoculation source, in addition to using new commercial inoculants.

### 4.2. Inoculation and Water Regime Effects

This study showed that the P uptake of pearl millet seedlings was more responsive to $I_1$ inoculant than rice seedlings (Figure 2a,b; Tables 5 and 6). The superior response of pearl millet to rice was also shown in the Andosol field experiment [26]. Inoculation improved shoot biomass from higher P uptake in maize at 10 weeks after sowing with 0, 10, and 50 g of $I_1$ in 2 kg of dry Andosol, and the relative effect was more noticeable under a low water regime due to the effect of P uptake on growth [21]. Under flooded (FL) conditions, the water regime may limit the infection rates by directly influencing aerobic AMF or increasing root aerenchyma [27]. This study found that rice did not increase phosphorus (P) uptake or shoot dry weight (SDW) through inoculation. It is possible that rice could absorb P directly due to a better-developed root system from the nursery soil.

The infection rates, measured as M%, A%, and V% through microscopic observation, were generally unaffected by the water regimes in our cell tray study. This finding was contrary to our expectation based on previous studies that indicated higher infection rates under water-limiting conditions than those with ample water supply (e.g., [5]). A similar negligible response in infection rates to water regimes was obtained in the same set of pearl millet and rice genotypes under Andosols in the field experiments [24,26]. The discrepancy between this result and the expectation might be related to the magnitude of changes in rhizosphere P availability when water availability changes, which needs further experimental proof. The small response in our cell tray experiments might be partially due to the low infection rates with the commercial soil media [28] and the small volume of soil in the cell tray. Water regimes did change the composition of AMF species in rice and pearl millet [24,25], but the overall infection rates did not dramatically change. Although this experiment cannot strongly prove it, water limitation (e.g., W25) may reduce carbon assimilation, make roots shorter and thinner, and induce complicated changes that influence mycorrhizal infection.

## 5. Conclusions

The new inoculants, Rootella ($I_2$ and $I_3$), with higher propagule numbers, showed a higher infection rate than the control seedlings in both rice and pearl millet, with the tendency for slower leaf development and no seedling growth enhancement. The popular Japanese inoculant Dr. Kinkon ($I_1$) had more positive effects on the root transversal area and shoot growth parameters than the control. The infection rates of all three inoculants were lower than the indigenous AMF from upland Andosol in rice and pearl millet, in which a higher infection rate led to a higher P uptake found in $millet_{954}$. The inoculant Dr. Kinkon increased the infection rate in pearl millet and rice but did not indicate interaction with water regimes. A higher infection rate led to a higher P uptake and shoot dry weight in pearl millet but not in rice with higher root length density. This study provided the significance of inoculants for seedling establishment and highlighted more mycorrhizal-mediated P uptake in pearl millet than in rice.

**Supplementary Materials:** The following supporting information can be downloaded at: https://www.mdpi.com/article/10.3390/agriculture15070753/s1, Figure S1: Mycorrhizal structures in infected roots dyed with trypan blue. Bar = 50 μm. Table S1. Effects of inoculation and genotype on mycorrhizal (M%), arbuscular (A%), and vesicular (V%) infection rates in (%), root length density (RLD, $cm/cm^3$), root transversal area (RTA, $\times 10^4$ $\mu m^2$), shoot P concentration (P%, %) and P uptake (mg/plant), shoot dry weight (SDW, g/plant), plant height (PH, cm), and leaf age (LA) with control (C) and inoculants Dr. Kinkon ($I_1$) and Rootella P ($I_2$) and F ($I_3$) in pearl millet (ICMB89111, $millet_{891}$; ICMB95444, $millet_{954}$) and rice (Koshihikari, $rice_k$; Togo4, $rice_t$) in 2020 (Exp. 1). Table S2. Effects of soil type and genotype on mycorrhizal (M%), arbuscular (A%), and vesicular (V%) infection rate (%), root length density (RLD, $cm/cm^3$), root transversal area (RTA, $\times 10^4$ $\mu m^2$), shoot P concentration (P%, %) and P uptake (mg/plant), shoot dry weight (SDW, g/plant), plant height (PH, cm), and leaf age (LA) with indigenous AMF from Andosol upland soil (UP) and paddy soil (PD) in pearl millet (ICMB89111, $millet_{891}$; ICMB95444, $millet_{954}$) and rice (Koshihikari, $rice_k$; Togo4, $rice_t$) in 2020 (Exp. 2). Table S3. Effects of water and genotype on mycorrhizal (M%), arbuscular (A%), and vesicular (V%) infection rate (%), root length density (RLD, $cm/cm^3$), root transversal area (RTA, $\times 10^4$ $\mu m^2$), shoot P concentration (P%, %) and P uptake (mg/plant), shoot dry weight (SDW, g/plant), and plant height (PH, cm) with inoculant Dr. Kinkon ($I_1$) in four water availabilities (flooded, FL; well irrigated, W100; 50% well irrigated, W50; 25% well irrigated, W25) in pearl millet (ICMB89111, $millet_{891}$; ICMB95444, $millet_{954}$) and rice (Koshihikari, $rice_k$; Togo4, $rice_t$) in 2020 (Exp. 3). Table S4. Effects of water, inoculation, and genotype on mycorrhizal (M%), arbuscular (A%), and vesicular (V%) infection rate (%), root length density (RLD, $cm/cm^3$), root transversal area (RTA, $\times 10^4$ $\mu m^2$), shoot P concentration (P%, %) and P uptake (mg/plant), shoot dry weight (SDW, g/plant), plant height (PH, cm), and tiller number (TN) of control (C) and $I_1$ (I) in 5 water regimes (flooded, FL; soil saturated, SS; well irrigated, W100; 50% well irrigated, W50; 25% well irrigated, W25) in pearl millet (ICMB89111, $millet_{891}$; ICMB95444, $millet_{954}$) and rice (Koshihikari, $rice_k$; Togo4, $rice_t$) in 2021 (Exp. 4).

**Author Contributions:** Conceptualization and methodology, A.K. and P.Y.; experimentation, data collection and analysis, P.Y.; supervision, project administration, and funding acquisition, A.K.; manuscript preparation, A.K. and P.Y. All authors have read and agreed to the published version of the manuscript.

**Funding:** This research was funded by KAKENHI C, grant number 20K05995 and 23K26885.

**Institutional Review Board Statement:** Not applicable.

**Data Availability Statement:** Publicly available datasets were analyzed in this study. These data can be found here: [https://1drv.ms/f/s!Ah0xdLQv16t7hoN17rcxYG3qN7LnmQ?e=Edj45F] (accessed on 19 March 2025).

**Acknowledgments:** We thank Daisuke Tsugama (The University of Tokyo, Japan) and D.K. Gupta (ICRISAT, India) for pearl millet seeds, and Kenji Jinushi (Research Institute of Rice Production & Technology Co., Ltd., Toyoake, Japan) for Togo4 hybrid rice seeds. We thank Hiromi Nakanishi (The University of Tokyo, Japan) for his guidance in mineral element analysis and Ryo Ohtomo (NARO, Japan) for his constructive comments on the manuscript. We also thank the technical staff of the Institute for Sustainable Agro-ecosystem Services (ISAS), The University of Tokyo, for greenhouse preparation.

**Conflicts of Interest:** The authors declare no conflicts of interest. The funders had no role in the design of the study; in the collection, analyses, or interpretation of data; in the writing of the manuscript; or in the decision to publish the results.

## References

1. Mbodj, D.; Effa-Effa, B.; Kane, A.; Manneh, B.; Gantet, P.; Laplaze, L.; Diedhiou, A.G.; Grondin, A. *Arbuscular mycorrhizal* symbiosis in rice: Establishment, environmental control and impact on plant growth and resistance to abiotic stresses. *Rhizosphere* **2018**, *8*, 12–26. [CrossRef]
2. Brundrett, M.; Bougher, N.; Dell, B.; Grove, T.; Malajczuk, N. *Working with Mycorrhizas in Forestry and Agriculture*; ACIAR: Canberra, Australia, 1996. Available online: https://www.researchgate.net/publication/227365112 (accessed on 30 December 2024).
3. Maiti, R.K.; Bidinger, F.R. *Growth and Development of the Pearl Millet Plant*; Research Bulletin; ICRISAT: Hyderabad, India, 1981; Volume 6, p. 14.
4. Solaiman, M.Z.; Hirata, H. Effect of *Arbuscular mycorrhizal* fungi inoculation of rice seedlings at the nursery stage upon performance in the paddy field and greenhouse. *Plant Soil* **1997**, *191*, 1–12. [CrossRef]
5. Augé, R.M. Water relations, drought and vesicular-*Arbuscular mycorrhizal* symbiosis. *Mycorrhiza* **2001**, *11*, 3–42. [CrossRef]
6. Salomon, M.J.; Demarmels, R.; Watts-Williams, S.J.; McLaughlin, M.J.; Kafle, A.; Ketelsen, C.; Soupir, A.; Bucking, H.; Cavagnaro, T.R.; van der Heijden, M.G.A. Global evaluation of commercial *Arbuscular mycorrhizal* inoculants under greenhouse and field conditions. *Appl. Soil Ecol.* **2022**, *169*, 104225. [CrossRef]
7. Redecker, D.; Schüßler, A.; Stockinger, H.; Stürmer, S.L.; Morton, J.B.; Walker, C. An evidence-based consensus for the classification of *Arbuscular mycorrhizal* fungi (Glomeromycota). *Mycorrhiza* **2013**, *23*, 515–531. [CrossRef]
8. Cavagnaro, T.R.; Smith, F.A.; Smith, S.E.; Jakobsen, I. Functional diversity in *Arbuscular mycorrhizas*: Exploitation of soil patches with different phosphate enrichment differs among fungal species. *Plant Cell Environ.* **2005**, *28*, 642–650. [CrossRef]
9. Solaiman, Z.; Senoo, K. Interactions between *Lotus japonicus* genotypes and *Arbuscular mycorrhizal* fungi. *J. Plant Interact.* **2005**, *1*, 179–186. [CrossRef]
10. Tawaraya, K.; Hirose, R.; Wagatsuma, T. Inoculation of *Arbuscular mycorrhizal* fungi can substantially reduce phosphate fertilizer application to *Allium fistulosum* L. and achieve marketable yield under field condition. *Biol. Fertil.* **2012**, *48*, 839–843. [CrossRef]
11. Hayashi, M.; Niwa, R.; Urashima, Y.; Suga, Y.; Sato, S.; Hirakawa, H.; Yoshida, S.; Ezawa, T.; Karasawa, T. Inoculum effect of *Arbuscular mycorrhizal* fungi on soybeans grown in long-term bare-fallowed field with low phosphate availability. *Soil Sci. Plant Nutr.* **2018**, *64*, 306–311. [CrossRef]
12. Solaiman, M.Z.; Hirata, H. Effects of indigenous *Arbuscular mycorrhizal* fungi in paddy fields on rice growth and N, P, K nutrition under different water regimes. *Soil Sci. Plant Nutr.* **1995**, *41*, 505–514. [CrossRef]
13. Marro, N.; Grilli, G.; Soteras, F.; Caccia, M.; Longo, S.; Cofré, N.; Borda, V.; Burni, M.; Janoušková, M.; Urcelay, C. The effects of *Arbuscular mycorrhizal* fungal species and taxonomic groups on stressed and unstressed plants: A global meta-analysis. *New Phytol.* **2022**, *235*, 320–332. [CrossRef]
14. Salomon, M.J.; Watts-Williams, S.J.; McLaughlin, M.J.; Bücking, H.; Singh, B.K.; Hutter, I.; Schneider, C.; Martin, F.M.; Vosatka, M.; Guo, L.; et al. Establishing a quality management framework for commercial inoculants containing *Arbuscular mycorrhizal* fungi. *iScience* **2022**, *25*, 104636. [CrossRef] [PubMed]
15. Niwa, R.; Koyama, T.; Sato, T.; Adachi, K.; Tawaraya, K.; Sato, S.; Hirakawa, H.; Yoshida, S.; Ezawa, T. Dissection of niche competition between introduced and indigenous *Arbuscular mycorrhizal* fungi with respect to soybean yield responses. *Sci. Rep.* **2018**, *8*, 7419. [CrossRef]
16. Nguyen, T.H.A.; Kamoshita, A.; Ramalingam, P. Genetic analysis of root vascular traits in a population from two temperate japonica rice ecotypes. *Plant Prod. Sci.* **2022**, *25*, 320–336. [CrossRef]
17. Phoura, Y.; Kamoshita, A.; Norisada, M.; Deshmukh, V. Eco-physiological evaluation of Stele Transversal Area 1 for rice root anatomy and shoot growth. *Plant Prod. Sci.* **2020**, *23*, 202–210. [CrossRef]
18. Phillips, J.M.; Hayman, D.S. Improved procedures for clearing roots and staining parasitic and vesicular-*Arbuscular mycorrhizal* fungi for rapid assessment of infection. *Trans. Br. Mycol. Soc.* **1970**, *55*, 158–161, IN16–IN18. [CrossRef]

19. Giovannetti, M.; Mosse, B. An evaluation of techniques for measuring vesicular-*Arbuscular mycorrhizal* infection in roots. *New Phytol.* **1980**, *84*, 489–500. [CrossRef]
20. Johnson, N.C.; Graham, J.H.; Smith, F.A. Functioning of mycorrhizal associations along the mutualism-parasitism continuum. *New Phytol.* **1997**, *135*, 575–585. [CrossRef]
21. Karasawa, T.; Ariharal, J.; Kasahara, Y. Effects of previous crops on *Arbuscular mycorrhizal* formation and growth of maize under various soil moisture conditions. *Soil Sci. Plant Nutr.* **2000**, *46*, 53–60. [CrossRef]
22. Wang, L.; Ning, C.; Pan, T.; Cai, K. Role of silica nanoparticles in abiotic and biotic stress tolerance in plants: A review. *Int. J. Mol. Sci.* **2022**, *23*, 1947. [CrossRef]
23. Ortas, I.; Akpinar, Ç. Response of maize genotypes to several *Mycorrhizal inoculums* in terms of plant growth, nutrient uptake and spore production. *J. Plant Nutr.* **2011**, *34*, 970–987. [CrossRef]
24. Phoura, Y. Effects of Water Regime and Inoculation with *Arbuscular mycorrhizal* Fungi on Growth and Mycorrhizal Communities of Rice and Pearl Millet. Ph.D. Thesis, University of Tokyo, Tokyo, Japan, 2024.
25. Ohtomo, R.; Kamoshita, A. Effects of water regime and inoculation with *Arbuscular mycorrhizal* fungi on mycorrhizal communities of roots of rice and pearl millet in upland and lowland fields. *Plant Root* **2024**, *18*, 10–21. [CrossRef]
26. Phoura, Y.; Ohtomo, R.; Nakanishi, H.; Kamoshita, A. Effects of *Arbuscular mycorrhizal* fungi inoculation on infection and growth of rice and pearl millet in upland fields with three water regimes. *Plant Prod. Sci.* **2023**, *26*, 350–363. [CrossRef]
27. Vallino, M.; Fiorilli, V.; Bonfante, P. Rice flooding negatively impacts root branching and *Arbuscular mycorrhizal* colonization, but not fungal viability. *Plant Cell Environ.* **2014**, *37*, 557–572. [CrossRef]
28. Corkidi, L.; Allen, E.B.; Merhaut, D.; Allen, M.F.; Downer, J.; Bohn, J.; Evans, M. Assessing the infectivity of commercial mycorrhizal inoculants in plant nursery conditions. *J. Environ. Hortic.* **2004**, *22*, 149–154. [CrossRef]

agriculture

MDPI

Article

# Characterization and Whole-Genome Sequencing of *Phytobacter palmae* WL65, a Plant Growth-Promoting Rhizobacterium First Isolated from Rice Rhizosphere Soil in Thailand

**Pisit Thamvithayakorn [1], Cherdchai Phosri [2], Vineet Vishal [3,4] and Nuttika Suwannasai [1,*]**

[1] Department of Microbiology, Faculty of Science, Srinakharinwirot University, 114 Sukhumvit 23, Watthana, Bangkok 10110, Thailand; pisit.thamvi@gmail.com

[2] Department of Biology, Faculty of Science, Nakhon Phanom University, Nakhon Phanom 48000, Thailand; cherd.phosri@npu.ac.th

[3] Department of Botany, Dr Shyama Prasad Mukherjee University, Ranchi 834008, Jharkhand, India; vineet.vishal73@gmail.com

[4] Department of Botany, Bangabasi Evening College, University of Calcutta, Kolkata 700009, West Bengal, India

* Correspondence: nuttika@g.swu.ac.th; Tel.: +66-8-67239088

**Abstract:** *Phytobacter palmae* WL65, isolated from the rice rhizosphere, was confirmed as *P. palmae* through whole-genome analysis. WL65 exhibited key plant growth-promoting (PGP) characteristics, including nitrogen fixation (*nifA*, *nifB*, *nifD*, *nifE*, *nifF*, *nifH*, *nifJ*, *nifK*, *nifL*, *nifS*, *nifU*, *nifW*, and *nifX*), phosphate solubilization (*pstA*, *pstB*, *pstC*, *pstS*, *phnC*, *phnD*, *phnE*, and *phnV*), siderophore production (*fhuA*, *fhuB*, *fhuC*, *fhuD*, *fhuF*, *feoA*, *feoB*, *feoC*, *acrA*, *acrB*, *acrE*, *acrR*, and *acrZ*), and phytohormone biosynthesis (*trpA*, *trpB*, *trpC*, *trpE*, *trpGD*, *trpR*, and *trpS*). WL65 also contains an enterobactin biosynthetic gene cluster, essential for iron acquisition and enhancing both bacterial survival and plant growth. This study provides the first genomic insights into the PGP characteristics of *P. palmae*. The application of WL65 in rice cultivation as a biostimulant resulted in effective root colonization, supported by biofilm formation genes (*pgaA*, *pgaB*, *pgaC*), which enhance bacterial adhesion. The treatment significantly improved rice growth, increasing plant height (5.8%), panicle length (10.2%), and seed yield (34.5%). Soil analysis revealed improved nutrient availability, including increased organic matter (21%), phosphorus (38.4%), potassium (29.8%), and calcium (27%) levels. These findings suggest that WL65 is a promising biofertilizer candidate for improving soil fertility and nutrient uptake in sustainable agriculture.

**Keywords:** *Phytobacter*; nitrogen-fixing bacteria; PGP; rice

## 1. Introduction

The genus *Phytobacter* (family Enterobacteriaceae, order Enterobacterales) was reclassified from *Enterobacter* and is recognized as a plant growth-promoting (PGP) bacteria. It currently comprises four accepted species: *P. diazotrophicus*, *P. palmae*, *P. massiliensis*, and *P. ursingii*. *P. diazotrophicus*, the type species, was first isolated as a nitrogen-fixing endophyte from *Oryza rufipogon* in China, highlighting its role in plant growth promotion [1]. *P. palmae*, isolated from oil palm leaves in Singapore, also exhibits nitrogen-fixing capabilities, further supporting its PGP potential [2]. In contrast, *P. ursingii*, isolated from clinical samples in Brazil, is distinguished by its ability to utilize L-sorbose and D-serine [3]. Its identification in healthcare-associated infections led to a revision of the genus description. The most recent addition, *P. massiliensis*, was reclassified from *Metakosakonia massiliensis* based on


https://doi.org/10.3390/agriculture15070707

genome relatedness and phylogenomic analyses [4]. Together, accumulating the genomic, biochemical, and ecological studies of *Phytobacter* species highlights their varied habitats, ranging from plant-associated environments to clinical situations, and their importance as PGP and potential biotechnological agents. Traditional bacterial identification methods, including morphology, biochemical tests, and 16S rRNA sequencing, often yield ambiguities, particularly among closely related taxa like *Phytobacter* species. Whole-genome sequencing (WGS) has emerged as a powerful tool to resolve taxonomic uncertainties, offering high-resolution genomic data, robust phylogenomic analyses, functional gene annotation, and key genetic feature identification.

Recently, Almuzara et al. [5] utilized WGS to investigate three clinical cases associated with *Phytobacter* spp. in Argentina. Through comparative genomic analyses, the isolates were identified as *P. diazotrophicus* and *P. ursingii*. The study highlighted the presence of antibiotic-resistant genes within these genomes, underscoring the clinical relevance of *Phytobacter* species and their potential role in healthcare-associated infections. This finding also demonstrates the versatility of WGS in linking genomic features to both ecological roles and pathogenic potential. In addition, a *Phytobacter* sp. isolate, WL65, was recently obtained from the rhizosphere soil of Jasmine rice (*Oryza sativa* L. cultivar KDML105) in Thailand. This isolate, initially identified by the morphological characters and 16S rRNA gene sequencing, exhibited plant growth-promoting properties and was subsequently used in combination with other plant growth-promoting rhizobacteria as a biostimulant to enhance rice growth [6]. However, the *Phytobacter* sp. WL65 had not yet been extensively characterized. Therefore, the objectives of this study were to conduct a comprehensive whole-genome sequencing analysis, including digital DNA-DNA hybridization, and to perform an extensive biochemical characterization to accurately identify the species and gain deeper insights into the functional potential of the isolates as plant growth promoters. Additionally, secondary metabolic pathways were predicted through genomic analysis, and the potential application of the isolate WL65 as a biofertilizer for rice cultivation was evaluated.

## 2. Materials and Methods

### 2.1. Bacterial Strains

*Phytobacter* sp. isolate WL65 was obtained from the Department of Microbiology, Faculty of Science, Srinakharinwirot University, Thailand. WL65 was originally isolated from the rhizosphere soil of organic Thai Jasmine rice (*Oryza sativa* L. cultivar KDML105) grown in Sakon Nakhon province, Thailand. WL65 was cultured on Luria–Bertani (LB) agar at 30 °C and preserved as a stock culture in 15% (*v*/*v*) glycerol at −80 °C for long-term storage.

### 2.2. Physiological and Biochemical Characteristics

The isolate WL65 was observed for phenotypic characteristics of bacterial colonies on a nitrogen-free medium, nutrient agar, and MacConkey agar after incubation at 30 °C and 37 °C for 24 h. Gram staining, endospore formation, and cellular shape were assessed. Catalase and oxidase activities were tested according to Aslanzadeh et al. [7]. Motility was determined using a semi-solid medium containing 0.4% (*w*/*v*) soft agar. The biochemical characteristics were analyzed using the API 20E test strip (bioMérieux, Marcy-l'Étoile, France) following the instructions of the manufacturer.

### 2.3. 16S rRNA Gene Sequences and Phylogenetic Analysis

The genomic DNA of isolate WL65 was extracted using the Favorprep™ Tissue genomic DNA extraction mini kit following the instructions of the manufacturer. The 16S rRNA gene was amplified using universal primers 27F and 1492R [8]. The purified PCR amplicon was sequenced by Macrogen Inc. (Seoul, Republic of Korea). The nucleotide sequence was submitted to the GenBank database under accession number PP326909.

The 16S rRNA gene sequence was carefully verified using the BioEdit software v7.2 [9] and aligned with sequences of the *Phytobacter* species and related genera obtained from the GenBank database using ClustalX [10]. A phylogenetic tree was constructed using the maximum likelihood (ML) method with a bootstrap analysis based on 1000 replications, implemented in MEGA X [11].

### 2.4. Genome Sequencing and Genome Annotation

Genomic DNA of the isolate WL65 from Section 2.3 was used to prepare the library using the Nextera XT DNA library preparation kit, and the extracted DNA was sequenced using the Illumina system through commercial services of Macrogen Inc. (Seoul, Republic of Korea). The quality of the obtained raw reads was checked using FastQC v0.11.9, followed by adapter and quality trimming with Trimmomatic v0.39 [12]. De novo assembly was performed using SPAdes v3.15.4 [13], and the assembly quality was evaluated with QUAST v5.2.0 [14]. Genome annotation and comprehensive genome analysis for the isolate WL65 were performed using SEED and Rapid Annotation using Subsystem Technology, RAST (https://rast.nmpdr.org) (accessed on 9 January 2025); Pathosystems Resource Integration Center, PATRIC (https://www.bv-brc.org) (accessed on 13 January 2025); and the Functional Annotation and Classification of Proteins of Prokaryotes, FACoP (http://facop.molgenrug.nl) (accessed on 18 January 2025) [15,16]. The circular genome map was constructed using the Proksee server (https://proksee.ca) (accessed on 18 January 2025), which generated a map such as coding sequences (CDS), transfer RNA (tRNAs), ribosomal RNA (rRNAs), and guanine-cytosine (GC) skew content. These data were integrated with the results from Prokka (Rapid Prokaryotic Genome Annotation) [17,18].

### 2.5. Genome Mining and Phylogenomic Analysis

To identify and classify secondary metabolite biosynthesis gene clusters and gene functions, the genomes of isolate WL65 were analyzed using AntiSMASH v8 (Antibiotics & Secondary Metabolite Analysis Shell) [19] (https://antismash.secondarymetabolites.org) (accessed on 9 January 2025). To identify the genes involved in antibiotic resistance and plasmids, the genomes data were analyzed using ResFinder v4.5.0 [20] (http://genepi.food.dtu.dk/resfinder) (accessed on 28 January 2025) and PlasmidFinder v2.1 [21] (https://cge.food.dtu.dk/services/PlasmidFinder) (accessed on 28 January 2025).

To compare the prokaryotic genome sequences, the average nucleotide identity (ANI) values between the genome of isolate WL65 and the reference strains from the GenomesDB database were calculated using JSpeciesWS [22] (https://jspecies.ribohost.com/jspeciesws/#home) (accessed on 8 January 2025). To assess bacterial strain identification at the genus or species level, pairwise genome sequence comparisons and phylogenomic tree construction were conducted using the Genome BLAST Distance Phylogeny (GBDP) approach and GGDC v4.0 (the Genome-to-Genome Distance Calculator) via the TYGS (Type Strain Genome Server) platform [23,24] (https://tygs.dsmz.de) (accessed on 29 January 2025). The genome was aligned against reference sequences using the progressive Mauve algorithm via Geneious Prime v.2025.0.3 [25].

### 2.6. *Phytobacter Palmae WL65 as a Plant Growth-Promoting Rhizobacterium for Rice Cultivation in Pot Experiments*

#### 2.6.1. Bacterial Colonization on Rice Roots

The rice (*Oryza sativa* L.) seeds cultivar RD79 was obtained from the Department of Agriculture, Thailand. The seeds were surface-sterilized by soaking them in 75% ethanol, followed by 0.6% sodium hypochlorite, and were then rinsed with sterile distilled water. The isolate WL65 was cultured in nutrient broth for 48 h at 30 °C with shaking at 150 rpm, and the bacterial suspension was adjusted to $1.0 \times 10^8$ CFU/mL in 0.85% (*v*/*v*) saline solution. Fifty rice seeds were immersed in 50 mL of bacterial suspension at 30 °C with shaking at 150 rpm for 24 h. The seeds were germinated in Petri dishes for 14 days at 30 °C in the dark. After that, four healthy seedlings were chosen, and the roots were immersed in 50 mL of WL65 suspension ($1.0 \times 10^8$ CFU/mL) for 1 h before transplanting to a minirhizotron containing 400 g of sterile soil. Four additional healthy seedlings treated with sterile distilled water were used as a negative control. The rice roots were collected at 4 weeks after transplanting (DAT) and carefully washed in sterile water to remove the soils. Then, the roots were air-dried and kept at 4 °C until processing. The root samples were prepared for bacterial colonization analysis by JEOL JSM-IT500HR scanning electron microscope (SEM) observation at the Scientific and Technological Research Equipment Centre, Chulalongkorn University (Thailand).

#### 2.6.2. Effects of *P. palmae* WL65 on Rice Cultivar RD79 in Pot Experiment

The pot experiment consisted of two treatments: WL65 and a negative control. The experiment was conducted in a greenhouse (4 × 3 × 2 m) at the Faculty of Science, Nakhon Phanom University, Nakhon Phanom province. The planting process and measurements followed the method described by Thamvithayakorn et al. [6]. The soil used for planting was collected from paddy fields in Nong Yart District, Nakhon Phanom province, sieved through a 5 mm mesh to remove stones, and mixed with organic fertilizer (Ngok-Ngam™, Ubon Ratchathani, Thailand) in a ratio of 1000:3 (*w/w*) before use.

Fifty RD79 rice seeds were surface-sterilized and germinated in soil for 21 days in the greenhouse. Eight seedlings with the same root and shoot length were selected, soaked in 100 mL of the treatment solutions (WL65 and control) for 1 h, and then transplanted into 10-inch pots containing 3 kg of the soil mixture. One seedling was placed in each pot. The pots were irrigated with tap water every two days.

Rice growth parameters, including the rice height, chlorophyll content index, number of panicles per tiller, panicle length, and number of seeds per panicle, were measured. Soil samples were collected and analyzed for physicochemical properties, including the pH, organic matter, total carbon, nitrogen, available phosphorus, potassium, calcium, and magnesium at Kasetsart University, Thailand.

### 2.7. *Statistical Analysis*

The experiments were performed in four replications and analyzed using the independent samples *t*-test in IBM SPSS Statistics v29.0.2. Data visualization was performed using RStudio v2024.12.0 + 467.

## 3. Results

### 3.1. *Biochemical Characteristics and 16S rRNA Gene Sequence Analysis*

Isolate WL65 was previously identified as a member of the genus *Phytobacter* based on the 16S rRNA gene sequence analysis in our previous study [6]. *Phytobacter* sp. WL65 is a Gram-negative, rod-shaped, motile bacterium that does not form spores. Its cell dimensions range from 0.62–0.82 μm in width to 1.62–2.96 μm in length. On a nitrogen-free medium and nutrient agar, the colonies appeared smooth, white, and circular with entire edges and measured 2–3 mm in diameter after 24 h of incubation at 30 °C. On the MacConkey agar, the colonies developed a pink color after incubation at 37 °C for 24 h. Biochemical characterization of isolate WL65 was performed using the API 20E system and compared to four reference *Phytobacter* species: *P. palmae* S29$^T$, *P. ursingii* ATCC 27989$^T$, *P. diazotrophicus* DSM 17806$^T$, and *P. massiliensis* JC163$^T$ (Table 1). The biochemical profile of isolate WL65 showed positive catalase activity but negative oxidase activity. It was capable of fermenting glucose, mannitol, sorbitol, rhamnose, saccharose, amygdalin, and arabinose but was unable to ferment inositol and melibiose. Additionally, it tested positive for citrate utilization and indole production.

**Table 1.** Biochemical characteristics of *Phytobacter* sp. WL65 and another four *Phytobacter* species based on the API20E analysis.

| Biochemical Characteristics | WL65 | 1 | 2 | 3 | 4 |
|---|---|---|---|---|---|
| Beta-galactosidase | + | + | + | + | + |
| Arginine dihydrolase | - | - | - | - | - |
| Lysine decarboxylase | - | - | - | - | - |
| Ornithine decarboxylase | - | - | - | - | + |
| Citrate utilization | + | + | + | + | - |
| $H_2S$ production | - | - | - | - | - |
| Urease | - | - | - | - | - |
| Tryptophan deaminase | - | + | - | - | - |
| Indole production | + | + | + | + | + |
| Acetoin production (Voges-Proskauer) | - | - | + | + | - |
| Gelatinase | - | - | - | - | + |
| Fermentation-oxidation of: | | | | | |
| Glucose | + | + | + | + | + |
| Mannitol | + | + | + | + | + |
| Inositol | - | - | - | - | - |
| Sorbitol | + | + | + | + | + |
| Rhamnose | + | + | + | + | + |
| Saccharose | + | + | + | + | - |
| Melibiose | - | - | - | - | - |
| Amygdalin | + | + | + | + | + |
| Arabinose | + | + | + | + | + |
| Cytochrome oxidase | - | - | - | - | - |

1: *P. palmae* S29$^T$; 2: *P. ursingii* ATCC 27989$^T$; 3: *P. diazotrophicus* DSM 17806$^T$; 4: *P. massiliensis* JC163$^T$. Data (1–4) from BacDive (https://bacdive.dsmz.de) (accessed on 28 December 2024).

A phylogenetic tree based on the 16S rRNA gene sequences of four *Phytobacter* species and related genera was constructed, as shown in Figure 1. The analysis identified three major clades corresponding to the genera *Enterobacter*, *Salmonella*, and *Phytobacter*. *Phytobacter* sp. WL65 clustered within the *Phytobacter* clade, showing a high sequence similarity of 99.70% to *P. palmae* S29$^T$. Comparatively, WL65 exhibited sequence similarities of 98.51%, 99.17%, and 98.44% to *P. diazotrophicus* DSM 17806$^T$, *P. ursingii* ATCC 27989$^T$, and *P. massiliensis* JC163$^T$, respectively. Although WL65 demonstrated more than 99% similarity to both *P. palmae* and *P. ursingii*, its biochemical characteristics were more closely aligned with those of *P. palmae*, except that WL65 was negative for tryptophan deaminase.

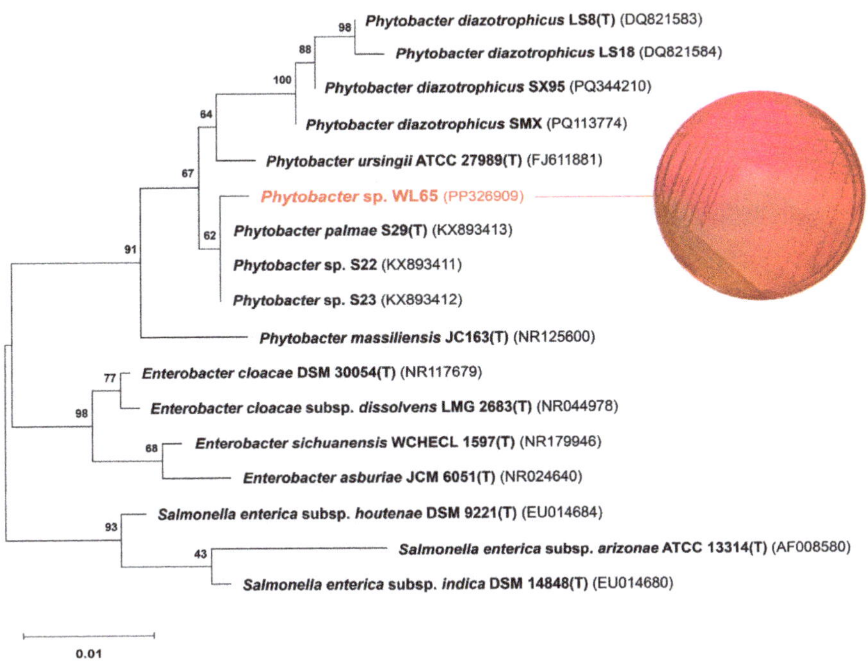


**Figure 1.** Phylogenetic relationship of *Phytobacter* sp. WL65 and related genera based on 16S rRNA gene sequences using the maximum likelihood method. The numbers on the branches were the percentages of the bootstrap support derived from 1000 replications. Bar was the nucleotide substitutions per nucleotide. The MacConkey agar exhibited the pink colonies of *Phytobacter* sp. WL65.

### *3.2. Genome Sequencing and Phylogenomics Analysis*

The draft genome sequence of isolate WL65 was analyzed, and its basic genome indices are summarized in Table 2, in comparison with the type strains *P. palmae* S29$^T$. The complete genome has been deposited at DDBJ/ENA/GenBank under the accession JBLQZN000000000, and a circular chromosome map was generated using PATRIC, as shown in Figure 2A. The genome size was 5,253,439 bp (5.2 Mb), with a G + C content of 52.51% and consisted of 33 contigs. The genome contained 5210 coding sequences (CDS), 58 tRNAs, 4 rRNAs, 347 subsystems, an N50 value of 340,721, and an L50 value of 5. No antibiotic resistance genes, transporters, drug target genes, or plasmids were identified in the WL65 genome based on the analyses using the respective databases. A genome annotation of isolate WL65, indicating subsystem features, is presented in Figure 2B. These features include key biological processes, including carbohydrates, amino acids and derivatives, protein metabolism, cofactors, vitamins, prosthetic groups, stress response, nitrogen metabolism, phosphorus metabolism, etc. (Table S1). Functional annotation was performed using the FaCoP system and identified the genes associated with essential cellular processes, including transcription (227 genes, 11.8%), inorganic ion transport and metabolism (216 genes, 11.3%), carbohydrate transport and metabolism (211 genes, 11%), amino acid transport and metabolism (173 genes, 9%), and various gene distributions (Table S2 and Figure S1).

A phylogenomic tree was constructed to include the isolate WL65 and related species, as shown in Figure 3. The results clearly demonstrated that isolate WL65 clustered with *P. palmae* S29$^T$ and PSU29, forming a distinct group separate from other species. Pairwise genome comparisons were conducted (Table S3), and the digital DNA-DNA hybridization (dDDH) values between the isolate WL65 and *P. palmae* S29$^T$ and PSU29 were 94.1% and

92.9%, respectively (Figure 4 and Figure S2). Isolate WL65 showed a 47.9% dDDH similarity to *P. ursingii* ATCC 27989$^T$ and 45.0% to *P. diazotrophicus* DSM 17806$^T$. Therefore, isolate WL65 was identified as *P. palmae* WL65.

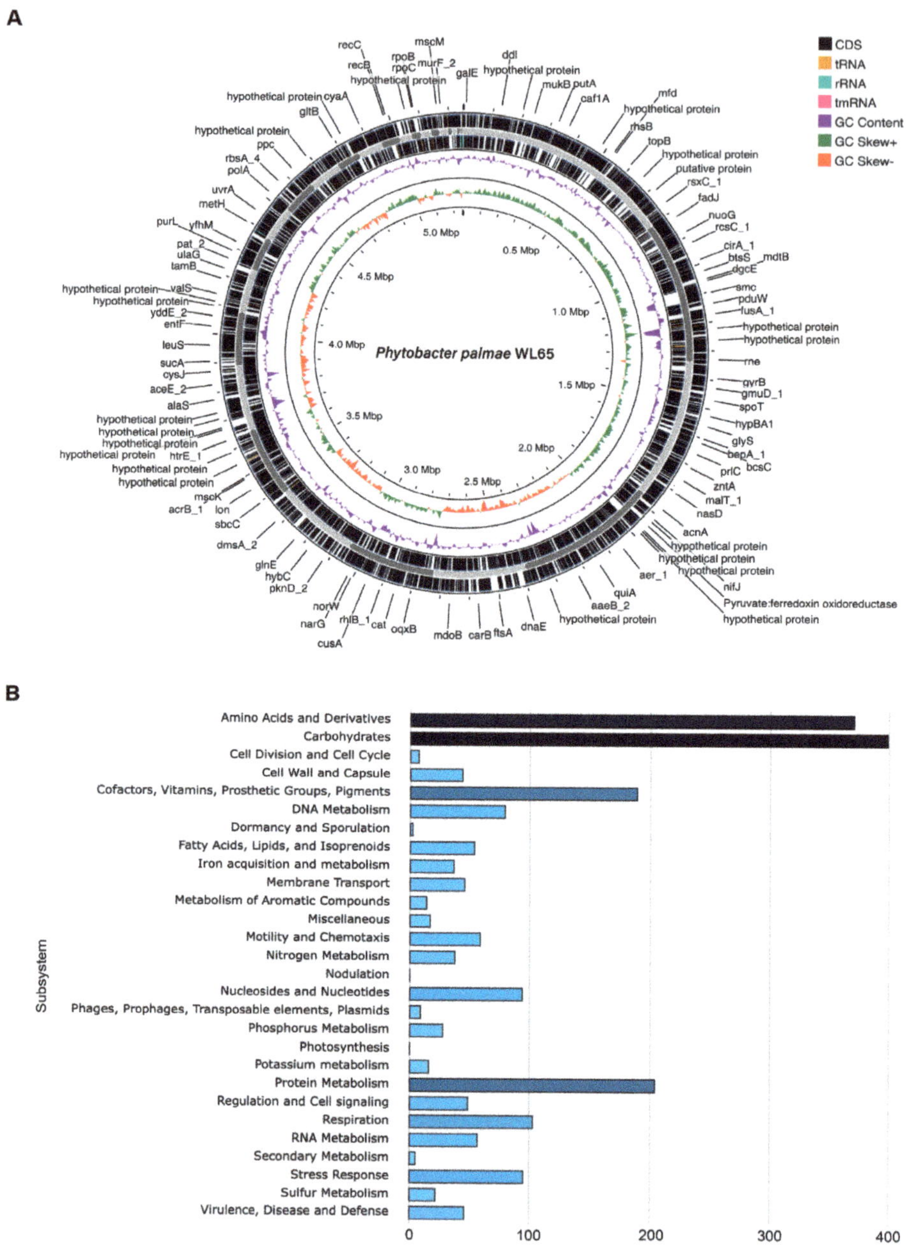


**Figure 2.** The draft genome of *P. palmae* WL65: (**A**) A circular genome map was generated using the Proksee server, showing coding sequences (CDS), transfer RNA (tRNA), ribosomal RNA (rRNA), transfer messenger RNA (tmRNA), guanine-cytosine (GC) content, and GC skew; (**B**) a bar chart illustrating the subsystem category distribution analyzed using the SEED and RAST servers.

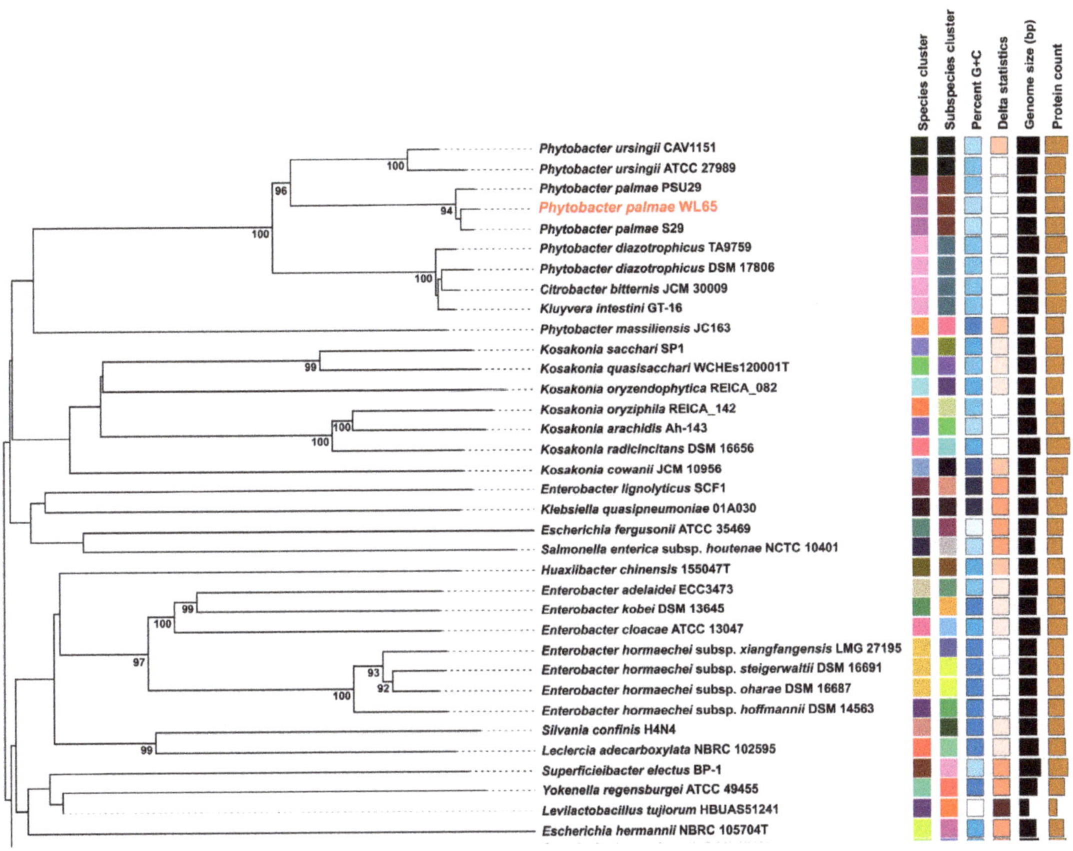


**Figure 3.** Genome BLAST distance phylogeny (GBDP) tree. The tree, inferred with FastME 2.1.6.1 from GBDP, provides the distances calculated from the genome sequences. The branch lengths are scaled in terms of the GBDP distance formula d5. The numbers above the branches are GBDP pseudo-bootstrap support values > 60% from 100 replications, with an average branch support of 61.7%. The tree was rooted at the midpoint.

**Table 2.** Overall statistical genome features of *P. palmae* WL65 and the type strain.

| Statistics | *P. palmae* WL65 | *P. palmae* S29$^{T}$ [1] |
|---|---|---|
| Genome size (bp) | 5,253,439 | 5,284,330 |
| G + C content (%) | 52.51 | 52.60 |
| Number of contigs | 33 | ND |
| Number of coding sequences | 5210 | ND |
| Number of tRNA | 58 | 70 |
| Number of rRNA | 4 | 13 |
| Number of subsystems | 347 | ND |
| N50 contig size (bp) | 340,721 | ND |
| L50 | 5 | ND |
| Plasmids | 0 | ND |
| Hypothetical proteins | 1071 | ND |
| Proteins with functional assignments | 4139 | ND |
| Proteins with EC number assignments | 1307 | ND |
| Proteins with GO assignments | 1080 | ND |
| Proteins with pathway assignments | 937 | ND |

[1] Type strain of *P. palmae* S29$^{T}$ [3].

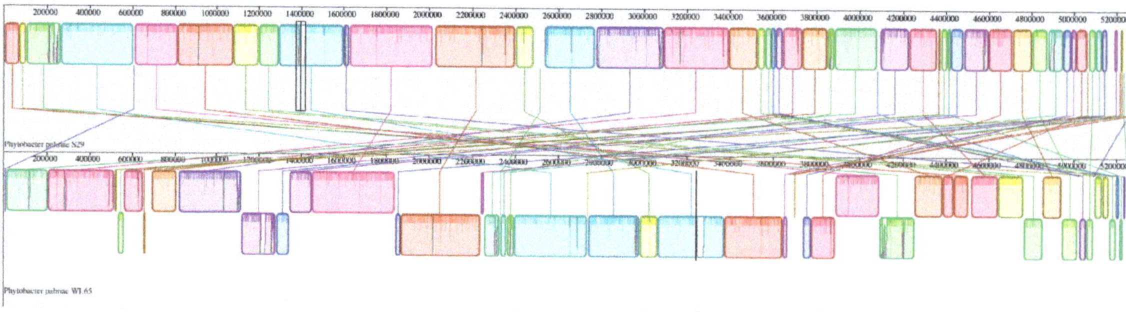


**Figure 4.** Pairwise comparison of the *P. palmae* WL65 (**bottom**) genome with the type strain *P. palmae* S29$^T$ (**top**) using the progressive MAUVE algorithm via Geneious Prime software v.2025.0.3. The same colors represent homologous locally collinear blocks between genomes, connected by lines.

### *3.3. Genome Analysis of P. palmae WL65 as Plant Growth-Promoting Rhizobacterium*

*P. palmae* WL65 exhibited key plant growth-promoting bacterium traits, including nitrogen fixation, phosphate solubilization, indole-3-acetic acid (IAA) production, siderophore production, and ACC deaminase activity, as previously reported by Thamvithayakorn et al. [6]. The genomic analysis revealed the presence of genes associated with these beneficial characteristics (Table 3 and Table S4). The nitrogen fixation genes *nifA*, *nifB*, *nifD*, *nifE*, *nifF*, *nifH*, *nifJ*, *nifK*, *nifL*, *nifS*, *nifU*, *nifW*, and *nifX* indicate the strain's capacity to convert atmospheric nitrogen ($N_2$) into ammonia ($NH_3$) by nitrogenase enzymes, thereby enhancing nitrogen availability for plants. Notably, *nifH*, *nifD*, and *nifK* encode the nitrogenase enzyme complex, which catalyzes nitrogen reduction, while *nifA* regulates *nif* gene expression. The presence of these genes suggests that *P. palmae* WL65 contributes significantly to nitrogen enrichment in the rhizosphere. The phosphate solubilization genes (*pstA*, *pstB*, *pstC*, *pstS*, *phnC*, *phnD*, *phnE*, and *phnV*) facilitate the uptake and mobilization of insoluble phosphate compounds, increasing phosphorus bioavailability. The presence of siderophore biosynthesis genes (*fhuA*, *fhuB*, *fhuC*, *fhuD*, *fhuF*, *feoA*, *feoB*, *feoC*, *acrA*, *acrB*, *acrE*, *acrR*, and *acrZ*) enables iron acquisition from the soil, improving plant iron uptake and alleviating iron deficiency stress. The potassium solubilization genes (*gcd1* and *gcd2*) were characterized in the genome of *P. palmae* WL65. These genes encode dehydrogenase, producing gluconic acid to solubilize potassium, enhancing plant nutrient uptake and growth. The genes involved in the tryptophan-dependent IAA biosynthetic pathway (*trpA*, *trpB*, *trpC*, *trpE*, *trpGD*, *trpR*, and *trpS*) contribute to plant growth promotion by stimulating root elongation and lateral root formation.

For plant defense, detoxification, and stress tolerance genes, several key genes were identified. Genes encoding ACC deaminase (*rimI*, *rimK*, *rimL*, *rimM*, *rimO*, *rimP*, and *rimJ*) facilitate the degradation of 1-aminocyclopropane-1-carboxylate (ACC), an ethylene precursor, helping to mitigate plant stress under drought and salinity conditions. Additionally, *P. palmae* WL65 possesses hydrogen sulfide ($H_2S$) biosynthesis genes (*cysA*–*cysZ*), which contribute to sulfur assimilation and plant defense. The *sodA*, *sodB*, and *sodC* genes encode the superoxide dismutase enzymes essential for oxidative stress protection, enhancing bacterial survival and plant vitality. Furthermore, the *osmB*, *osmC*, *osmE*, *osmV*, *osmW*, *osmX*, *osmY*, and *oxyR* genes encode peroxidases and oxidative stress-related proteins, supporting stress tolerance, detoxification, and plant growth promotion. Additionally, the *phzF* gene, responsible for phenazine production, enhances biocontrol and stress tolerance, making *P. palmae* WL65 more effective in sustainable agriculture.

**Table 3.** Genes involved in plant growth-promoting traits based on RAST and Prokka.

| PGP Traits | Genes | Functions |
|---|---|---|
| Nitrogen fixation | *nifA, nifB, nifD, nifE, nifF, nifH, nifJ, nifK, nifL, nifS, nifU, nifW, nifX* | Nitrogenase (molybdenum-iron)-specific transcriptional regulator; nitrogenase molybdenum-iron protein alpha chain; nitrogenase FeMo-cofactor carrier protein, etc. |
| Phosphate solubilization | *pstA, pstB, pstC, pstS, phnC, phnD, phnE, phnV* | Phosphate transport system permease protein; phosphate import ATP-binding protein; phosphate-binding protein; etc. |
| Siderophore production | *fhuA, fhuB, fhuC, fhuD, fhuF, feoA, feoB, feoC, acrA, acrB, acrE, acrR, acrZ* | Ferrichrome outer membrane transporter/phage receptor; Iron(3+)-hydroxamate import system permease protein; etc. |
| Indole-3-acetic acid production | *trpA, trpB, trpC, trpE, trpGD, trpR, trpS* | Tryptophan synthase alpha chain; tryptophan synthase beta chain; tryptophan biosynthesis protein; anthranilate phosphoribosyltransferase; etc. |
| Potassium solubilization | *gcd_1, gcd_2* | Glucose dehydrogenase, PQQ-dependent; quinoprotein glucose dehydrogenase |
| ACC deaminase activity | *rimI, rimK, rimL, rimM, rimO, rimP, rimJ* | ribosomal protein S18 alanine N-acetyltransferase; ribosomal maturation factor; etc. |
| Hydrogen sulfide production | *cysA, cysB, cysC, cysD, cysE, cysG, cysH, cysI, cysJ, cysK, cysL, cysM, cysN, cysP, cysQ, cysS, cysT, cysW, cysZ* | Adenylylsulfate kinase; sulfate adenylyltransferase subunit 1/2; thioredoxin-dependent 5′-adenylylsulfate reductase; etc. |
| Quorum sensing | *luxR, luxS* | Two-component transcriptional response regulator; S-ribosylhomocysteine lyase |
| Superoxide dismutase | *sodA, sodB, sodC* | Superoxide dismutase [Mn/Fe/Cu-Zn] |
| Biofilm production | *pgaA, pgaB, pgaC* | biofilm PGA outer membrane secretin; biofilm PGA synthesis deacetylase; biofilm PGA synthesis N-glycosyltransferase |
| Peroxidases | *osmB, osmC, osmE, osmV, osmW, osmX, osmY, oxyR* | Peroxiredoxin; osmotically-inducible lipoprotein; osmoprotectant import permease protein; etc. |
| Acetoin and butanediol synthesis | *budC, alsT, poxB* | Diacetyl reductase [(S)-acetoin forming]; amino acid carrier protein; pyruvate dehydrogenase |
| Phenazine production | *phzF* | Trans-2,3-dihydro-3-hydroxyanthranilate isomerase; phenazine biosynthesis protein |
| Glycine-betaine production | *proA, proB, proC, proP, proS, proX, proQ, proV, proW, proY, soxR, soxS* | L-Proline/glycine-betaine transporter; glycine-betaine/proline betaine transport system ATP-binding protein, etc. |
| Tryptophanase | *tnaA* | Tryptophanase |
| OqxAB efflux pump | *oqxB* | OqxAB efflux pump |
| GABA production | *gabR* | HTH-type transcriptional regulatory protein |

The colonization and quorum sensing-related genes were also identified. Genes involved in biofilm formation (*pgaA, pgaB, pgaC*) contribute to bacterial adhesion, enhancing rhizosphere colonization and plant-microbe interactions. Additionally, the strain possesses genes associated with acetoin and butanediol biosynthesis (*budC, alsT*, and *poxB*), GABA production (*gabR*), glycine-betaine synthesis (*proA, proB, proC, proP, proS, proX, proQ, proV, proW, proY, soxR*, and *soxS*), and quorum sensing (*luxR* and *luxS*), which further support plant resilience and microbial communication. Notably, a significant proportion of hypothetical proteins were identified, suggesting the presence of novel functional elements that warrant further investigation. Importantly, no plasmids carrying antibiotic-resistant genes were detected, highlighting the genomic safety of *P. palmae* WL65 for agricultural applications.

### 3.4. Genome Analysis of Secondary Metabolism of P. palmae WL65 Using AntiSMASH

Antimicrobial activity is one of the key traits of plant growth-promoting bacterium for plant growth, particularly in suppressing phytopathogens. The genome analysis of *P. palmae* WL65 using AntiSMASH identified six gene clusters that are involved in the biosynthesis of secondary metabolites (Table 4 and Figure 5). Among these, the azole-containing RiPP cluster encodes ribosomal synthesized and post-translationally modified peptides (RiPPs) that are characterized by azole heterocycles. These compounds often exhibit antimicrobial, antifungal, or anticancer properties. The genome also contains RiPP-like gene clusters, which have been identified through genome mining and untargeted metabolomics, potentially encoding novel bioactive compounds with pharmaceutical applications, including antimicrobial and anticancer activities. Additionally, a terpene-precursor biosynthesis cluster was identified. Terpene precursors are metabolic intermediates produced from primary metabolism that form the basis for terpene production. These molecules undergo enzymatic modifications, such as cyclization, oxidation, and glycosylation, leading to the production of a diverse array of bioactive terpenoids commonly found in plants, fungi, and microorganisms. Notably, the genome of *P. palmae* WL65 contains a non-ribosomal peptide (NRP) metallophore biosynthetic cluster, which exhibits high similarity to the enterobactin biosynthetic pathway. Enterobactin, also known as enterochelin, is a high-affinity siderophore essential for iron acquisition. It facilitates ferric iron ($Fe^{3+}$) scavenging from the environment and its transport into bacterial cells, enhancing bacterial survival and competitiveness under iron-limiting conditions.

**Table 4.** Secondary metabolite biosynthetic gene clusters (BGCs) of *P. palmae* WL65 identified using AntiSMASH.

| Region | BGCs Type | From (nt) | To (nt) | Most Similar Known Cluster |
|---|---|---|---|---|
| 1.1 | azole-containing-RiPP | 197,800 | 224,234 | - |
| 1.2 | RiPP-like | 671,131 | 682,162 | - |
| 8.1 | terpene-precursor | 42,358 | 63,257 | - |
| 10.1 | NRP-metallophore, NRPS | 105,627 | 159,362 | Enterobactin, NRP |
| 14.1 | RiPP-like | 1 | 8289 | - |
| 16.1 | terpene-precursor | 1676 | 22,647 | - |

### 3.5. P. palmae WL65 as a Plant Growth-Promoting Rhizobacterium (PGPR) for Rice Cultivation in Pot Experiments

*P. palmae* WL65 exhibited key PGPR properties, as supported by the genome analysis. This strain has the potential as a PGPR for rice cultivation, given its origin in the rice rhizosphere. The ability of *P. palmae* WL65 to colonize rice roots was assessed using scanning electron microscopy (SEM), which revealed a high density of bacterial cells of similar size and shape surrounding the roots after four weeks of seedling cultivation, compared to the uninoculated control (Figure 6). This finding confirms the strain's ability to effectively colonize plant roots, a critical trait for PGPR function capability. The enhanced root colonization observed may contribute to improved nutrient uptake and plant growth.

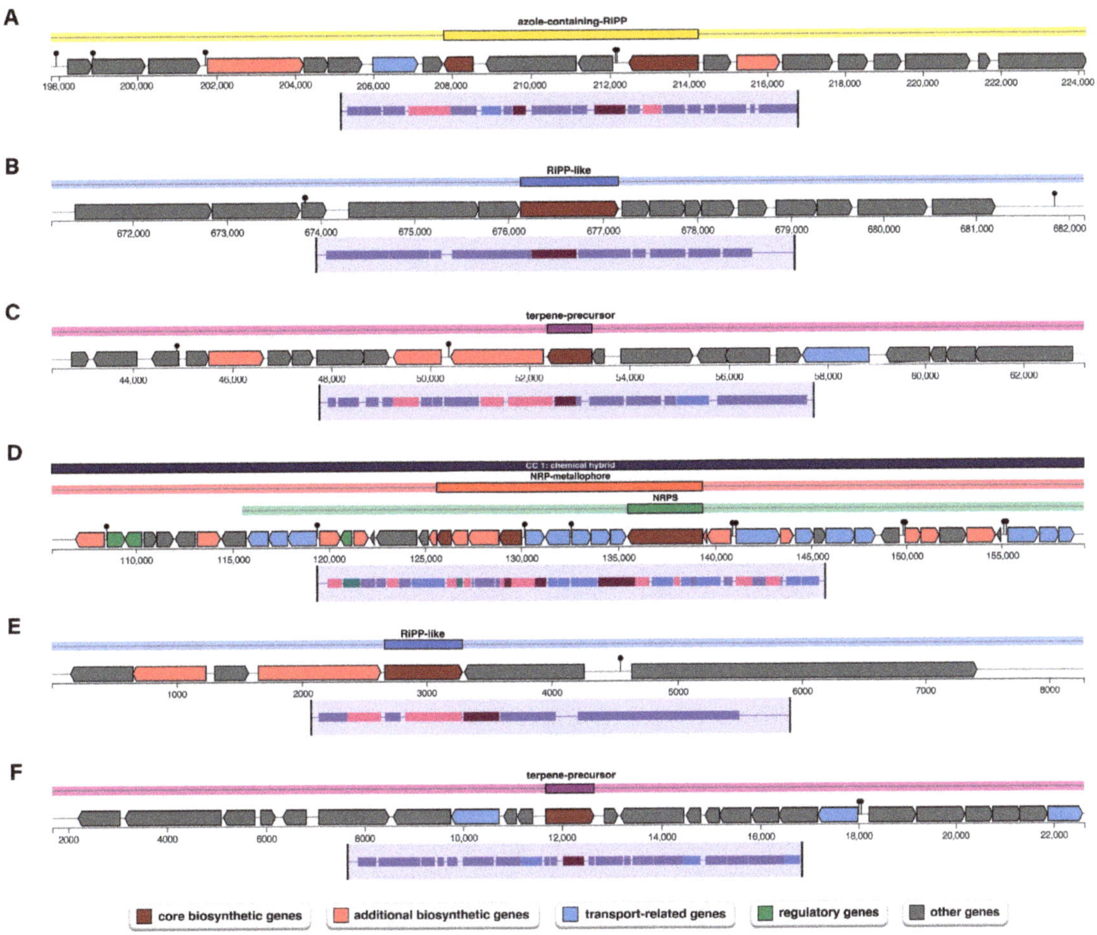


**Figure 5.** Biosynthetic gene clusters (BGCs) encoding secondary metabolites in the *P. palmae* WL65 genome: (**A**) azole-containing-RiPP; (**B**) RiPP-like; (**C**) terpene-precursor; (**D**) NRP-metallophore, NRPS; (**E**) RiPP-like; and (**F**) terpene-precursor.

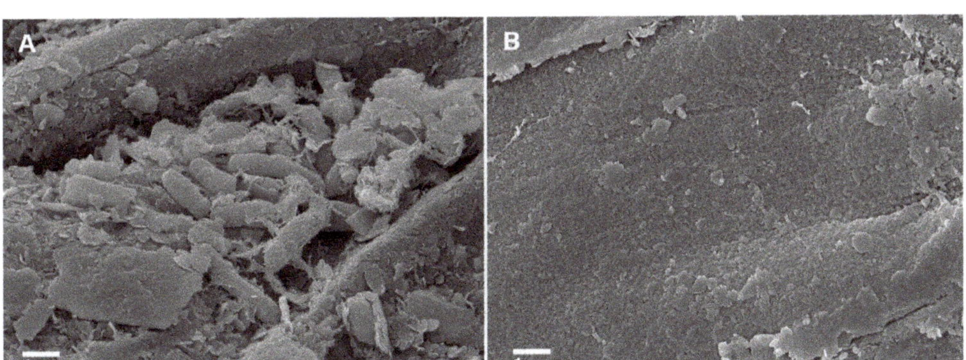


**Figure 6.** SEM micrographs of PGPR colonization on rice roots (**A**) *P. palmae* WL65 and (**B**) uninoculated control.

Further, *P. palmae* WL65 was evaluated as a PGPR in a pot experiment, where its impact on rice growth and soil physicochemical properties was assessed (Figures 7 and S3 and Table S5). Rice plants inoculated with WL65 exhibited a significantly greater plant height (91.67 ± 0.58 cm) compared to the control (86.67 ± 2.08 cm), representing a 5.8% increase. Additionally, the panicle length (21.02 ± 0.77 cm) and the number of seeds per panicle (68.17 ± 4.54 seeds) were significantly higher than those in the control (19.07 ± 0.92 cm and 50.67 ± 9.29 seeds, respectively), corresponding to increases of 10.2% and 34.5%. However, the chlorophyll content index and the number of panicles per tiller did not differ significantly between the WL65 treatment and the control. The chlorophyll content indices were 39.22 ± 0.36 for WL65 and 38.65 ± 1.68 for the control. Similarly, the number of panicles per tiller was 7.67 ± 0.58 panicles in the WL65 treatment and 8.33 ± 0.58 panicles in the control. The physicochemical analysis of the soil samples post-harvest (Table 5) indicated that the organic matter, available phosphorus, potassium, calcium, and magnesium levels were higher in the WL65 treatment compared to the control, suggesting an improvement in soil fertility that was associated with bacterial inoculation.

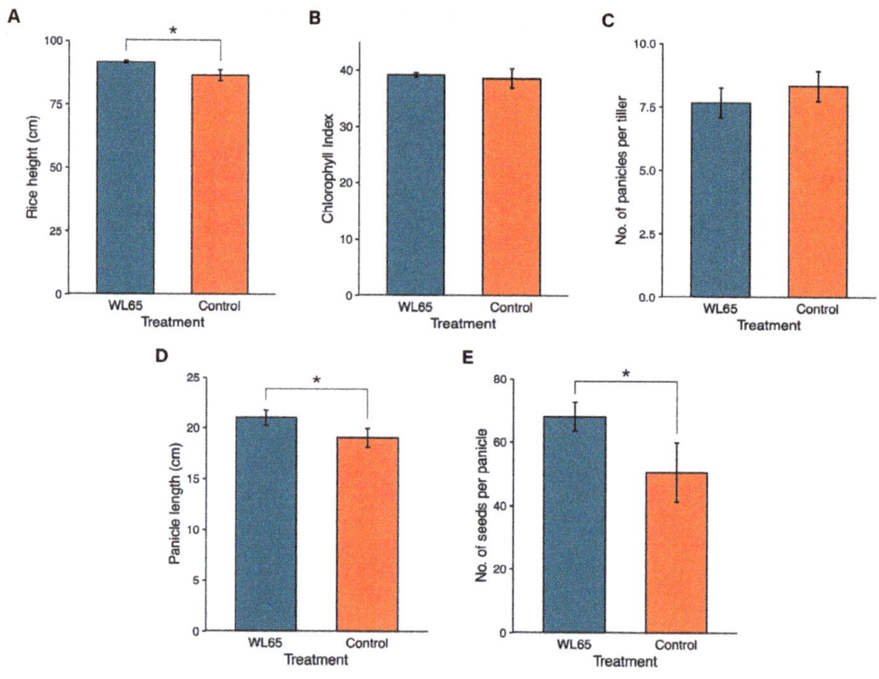


**Figure 7.** Effects of *P. palmae* WL65 on rice growth parameters: (**A**) rice height; (**B**) chlorophyll index; (**C**) number of panicles per tiller; (**D**) panicle length compared to (**E**) the uninoculated control. * Indicates a statistically significant difference at $p \leq 0.05$ (independent samples $t$-test).

**Table 5.** Physicochemical properties of soil samples from the pot experiment after harvest.

| Soil Properties | WL65 | Control |
|---|---|---|
| Soil texture | Sandy loam | Sandy loam |
| pH (1:1) | 6.08 | 5.99 |
| Organic matter (%) | 3.40 | 2.81 |
| EC 1:5 (dS m$^{-1}$) | 0.04 | 0.02 |
| Carbon (%) | 2.66 | 2.67 |

Table 5. Cont.

| Soil Properties | WL65 | Control |
|---|---|---|
| Nitrogen (%) | 0.26 | 0.26 |
| Available phosphorus (mg kg$^{-1}$) | 13.99 | 10.11 |
| Potassium (mg kg$^{-1}$) | 12.12 | 9.34 |
| Calcium (mg kg$^{-1}$) | 365.80 | 288.00 |
| Magnesium (mg kg$^{-1}$) | 27.94 | 20.04 |

## 4. Discussion

*Phytobacter* is a nitrogen-fixing bacterial genus consisting of only four species. In a recent study, we isolated rhizobacterial strains from the rhizosphere soils of rice (*Oryza sativa* L.) in northeast Thailand [6]. One isolate, *Phytobacter* sp. WL65, exhibiting outstanding plant growth-promoting properties, was selected for species identification. The biochemical and genomic characterization of WL65 confirmed its classification as *P. palmae*. This classification was supported by pairwise genome analysis and dDDH, which revealed a 94.1% genome similarity to the type strain *P. palmae* S29$^{T}$ and 92.9% similarity to *P. palmae* PSU29 [3,26]. Notably, these are the only two available genomic references for *P. palmae* in public databases. Interestingly, the biochemical profile of WL65 differed from that of *P. palmae* S29$^{T}$, particularly in tryptophan metabolism. While S29 exhibited tryptophan deaminase activity, WL65 tested negative for this enzyme. Tryptophan deaminase, encoded by the *tdcB* gene, catalyzes the deamination of tryptophan, leading to the production of indole-3-pyruvate and other metabolites, which also contribute to indole production. Despite the absence of *tdcB*, WL65 exhibited indole production, suggesting the occurrence of an alternative metabolic pathway for indole production. The genome analysis revealed that WL65 possesses *tnaA*, encoding tryptophanase, an enzyme that directly converts tryptophan into indole, ammonia, and pyruvate. This finding suggests that WL65 relies on TnaA-mediated tryptophan degradation rather than the TdcB-dependent pathway observed in S29 [27]. The variation in tryptophan metabolism between these strains may be influenced by environmental conditions and horizontal gene transfer. Notably, *tnaA* is more commonly expressed under aerobic conditions, whereas *tdcB* is typically induced under anaerobic or fermentative conditions [27].

Although *P. palmae* S29$^{T}$ was originally isolated as an endophyte from oil palm tissues in Singapore, exhibiting strong nitrogen-fixing ability [3], our study identified WL65 from rice rhizosphere soil in Thailand, where it demonstrated nitrogen fixation along with additional plant growth-promoting properties [6]. This suggests that *P. palmae* is not plant-specific but rather a versatile species capable of colonizing different plant hosts. Few *P. palmae* isolates have been reported, including two nitrogen-fixing strains identified in the present study and the type strain. However, more recently, *P. palmae* PSU29 was isolated from a clinical sample in Southern Thailand [26]. The genome analysis of PSU29 revealed the presence of the *oqxA* and *oqxB* genes, which encode the OqxAB efflux pump. The *oqxAB* operon is commonly found in Enterobacteriaceae, particularly in *Escherichia coli*, *Salmonella enterica*, and *Enterobacter* spp. [28,29]. Interestingly, the genome analysis of *P. palmae* WL65 revealed the presence of only the *oqxB* gene, while *oqxA* was absent. This suggests a possible genomic variation or rearrangement in this strain. Unlike PSU29, WL65 does not have any mobile plasmids associated with antibiotic resistance, indicating that it lacks horizontally acquired resistance elements. Additionally, while PSU29 was identified in a clinical sample, WL65 was isolated from the rice rhizosphere, highlighting the environmental diversity of *P. palmae* and its potential adaptation to different ecological niches. However, the genome analysis of WL65 revealed the presence of key genes associated with plant growth-promoting (PGP) traits, including nitrogen fixation, phosphate solubilization, IAA biosynthesis, and siderophore production. These findings confirm the PGPR potential of WL65, as previously described by Thamvithayakorn et al. [6]. In addition to these well-established PGP traits, WL65 has several genes essential for bacterial survival, plant colonization, stress tolerance, and pathogen suppression, further supporting its role in sustainable agriculture. Notably, this study represents the first genomic analysis of PGPR-associated traits in *P. palmae*, providing insights into its secondary metabolite biosynthetic potential. One of the key findings was the presence of an enterobactin biosynthetic gene cluster. Enterobactin is a well-characterized siderophore that plays a crucial role in iron acquisition, a vital function for both bacterial competitiveness in the rhizosphere and plant health [30,31]. The ability of WL65 to produce siderophores further reinforces its potential as a beneficial PGPR strain capable of enhancing plant growth and resilience under nutrient-limited conditions. These findings support the potential of *P. palmae* WL65 as a promising candidate for biofertilizer development, contributing to sustainable and eco-friendly agricultural practices.

Thailand is one of the largest rice producers, having a rice diversity of over 100 rice varieties, including the hybrid rice RD79, developed by the Rice Department (RD). RD79 is among the most popular rice varieties in Thailand due to its high yield and resistance to rice blast disease [32]. Therefore, WL65 was selected as a potential PGPR for RD79 cultivation. The results demonstrated successful bacterial colonization on rice roots compared to the control. Root colonization represents one of the key characteristics of PGPR, as it allows beneficial bacteria to form a stable association with plants and thus exerts their beneficial growth-promoting effects. The colonization ability of WL65 might be attributed to biofilm-associated genes (*pgaA*, *pgaB*, and *pgaC*), which are essential for bacterial adhesion to plant roots. The *pgaABC* operon encodes proteins that are involved in the biosynthesis of the extracellular polymeric substance poly-β-1,6-N-acetyl-D-glucosamine (PNAG), which is important for surface attachment, biofilm formation, and structural integrity. More specifically, *pgaC* codes a glycosyltransferase-linking PNAG, *pgaA* codes for an outer membrane transporter of PNAG, and *pgaB* deacetylates PNAG to increase its adhesive properties [33,34]. Overall, the synergistic action of these functions helps to promote bacterial clustering, persistence in the rhizosphere, and sustain colonization of plant roots, which are all meant to enable effective PGPR activity. Additionally, biofilm formation enhances bacterial survival by increasing resistance to environmental stressors, promoting root surface attachment, stabilizing bacterial populations in the rhizosphere, and improving nutrient exchange [35]. The root colonization of WL65 in the rice rhizosphere was confirmed by SEM analysis. Notably, biofilm-forming PGPR such as *Brucella* sp. and *Lysinibacillus macrolides* have been shown to enhance wheat (*Triticum aestivum* L.) growth and productivity, leading to increases in plant height and grain yields by 16.7% and 17.5%, respectively, compared to the control [36]. The application of WL65 significantly improved the rice growth parameters, including the plant height, panicle length, and the number of seeds per panicle, compared to the control. These growth enhancements are likely associated with the production of phytohormones such as IAA, a key auxin involved in root and shoot development [37]. Additionally, the soil physicochemical properties after harvest indicated improved soil health in the WL65 treatment compared to the control. The higher organic matter content observed in the treated soil may result from the enhanced nutrient cycling driven by beneficial bacterial activity in the rhizosphere. Furthermore, the increased availability of phosphorus and potassium in the soil confirms the phosphate and potassium solubilization capabilities of WL65, supporting its role in enhancing soil fertility and plant nutrient uptake. Notably, calcium levels in the WL65 treatment increased by 27% compared to the control, contributing to soil organic carbon sequestration and reducing $CO_2$ release [38]. Calcium also plays a crucial role in bacterial adhesion and modulating

PGPR-induced defense responses against environmental stress [39]. These results indicate that *P. palmae* WL65 is a novel PGPR candidate with strong potential for rice farming due to its plant growth-promoting activity. Given that most PGPRs are not plant-specific, future research should explore their application in other crops. Moreover, WL65 can serve as an eco-friendly alternative to conventional biofertilizers, supporting plant growth and biomodification while contributing to sustainable agriculture.

## 5. Conclusions

This study provides the first genomic characterization of *P. palmae* WL65 as a PGPR. Whole-genome analysis confirmed its identity, revealing genes for nitrogen fixation, phosphate solubilization, IAA biosynthesis, siderophore production, and biofilm formation. Despite its genomic similarity to *P. palmae* S29$^{T}$, WL65 exhibited metabolic differences in tryptophan metabolism. Its application in rice cultivation enhanced root colonization, plant growth, and yield while improving soil fertility. These findings highlight the potential of WL65 as a biofertilizer, contributing to sustainable agriculture and expanding our understanding of PGPR mechanisms for eco-friendly crop production.

**Supplementary Materials:** The following supporting information can be downloaded at: https://www.mdpi.com/article/10.3390/agriculture15070707/s1, Table S1: Subsystem of *P. palmae* WL65 distribution from SEED and Rapid Annotation using Subsystem Technology (RAST) servers; Table S2: Clusters of orthologous genes of *P. palmae* WL65 distribution using Functional Annotation and Classification of Proteins of Prokaryotes (FACoP); Figure S1: Clusters of orthologous genes distribution of *P. palmae* WL65; Table S3: Pairwise digital DNA-DNA hybridization (dDDH) comparisons of *P. palmae* WL65 genome with type-strain genomes; Figure S2: Pairwise comparisons of *P. palmae* WL65 genome with *P. palmae* PSU29 using the progressive mauve algorithm; Table S4: Genes involved in plant growth-promoting traits using RAST and Rapid Prokaryotic Genome Annotation (Prokka); Table S5: Effects of *P. palmae* WL65 on rice growth parameters in a pot experiment under greenhouse conditions; Figure S3: Rice pot experiment under greenhouse conditions showing WL65 treatment; T1 (left) and uninoculated control; T2 (right).

**Author Contributions:** Conceptualization, P.T., C.P. and N.S.; methodology, P.T. and N.S.; software, P.T.; validation, C.P., V.V. and N.S.; formal analysis, P.T. and N.S.; investigation, P.T.; resources, C.P.; data curation, P.T., C.P. and N.S.; writing—original draft preparation, P.T. and N.S.; writing—review and editing, C.P. and V.V.; supervision, N.S.; funding acquisition, P.T. and N.S. All authors have read and agreed to the published version of the manuscript.

**Funding:** This study was supported by the National Research Council of Thailand (NRCT) and ADAMA (THAILAND) LIMITED: grant no. N41A650169.

**Institutional Review Board Statement:** Not applicable.

**Data Availability Statement:** Data are contained within the article and Supplementary Materials.

**Acknowledgments:** We would like to thank Pravech Ajawatanawong, Mahidol University, Thailand, for his help with whole-genome sequencing.

**Conflicts of Interest:** The authors declare no conflicts of interest.

## References

1. Zhang, G.X.; Peng, G.X.; Wang, E.T.; Yan, H.; Yuan, Q.H.; Zhang, W.; Lou, X.; Wu, H.; Tan, Z.Y. Diverse endophytic nitrogen-fixing bacteria isolated from wild rice *Oryza rufipogon* and description of *Phytobacter diazotrophicus* gen. nov. sp. nov. *Arch. Microbiol.* **2008**, *189*, 431–439. [CrossRef] [PubMed]
2. Madhaiyan, M.; Saravanan, V.S.; Blom, J.; Smits, T.H.; Rezzonico, F.; Kim, S.-J.; Weon, H.-Y.; Kwon, S.-W.; Whitman, W.B.; Ji, L. *Phytobacter palmae* sp. nov., a novel endophytic, N2 fixing, plant growth promoting Gamma proteobacterium isolated from oil palm (*Elaeis guineensis* Jacq.). *Int. J. Syst. Evol. Microbiol.* **2020**, *70*, 841–848. [CrossRef]

3. Pillonetto, M.; Arend, L.N.; Faoro, H.; D'Espindula, H.R.; Blom, J.; Smits, T.H.; Mira, M.T.; Rezzonico, F. Emended description of the genus *Phytobacter*, its type species *Phytobacter diazotrophicus* (Zhang 2008) and description of *Phytobacter ursingii* sp. nov. *Int. J. Syst. Evol. Microbiol.* **2018**, *68*, 176–184. [CrossRef] [PubMed]
4. Ma, Y.; Yao, R.; Li, Y.; Wu, X.; Li, S.; An, Q. Proposal for unification of the genus *Metakosakonia* and the genus *Phytobacter* to a single genus *Phytobacter* and reclassification of *Metakosakonia massiliensis* as *Phytobacter massiliensis* comb. nov. *Curr. Microbiol.* **2020**, *77*, 1945–1954. [CrossRef]
5. Almuzara, M.; Cittadini, R.; Traglia, G.; Haim, M.S.; De Belder, D.; Alvarez, C.; O'Connor, Z.; Ocampo, C.V.; Barberis, C.; Prieto, M. *Phytobacter* spp: The emergence of a new genus of healthcare-associated Enterobacterales encoding carbapenemases in Argentina: A case series. *Infect. Prev. Pract.* **2024**, *6*, 100379. [CrossRef]
6. Thamvithayakorn, P.; Phosri, C.; Robinson-Boyer, L.; Limnonthakul, P.; Doonan, J.H.; Suwannasai, N. The synergistic impact of a novel plant growth-promoting rhizobacterial consortium and *Ascophyllum nodosum* seaweed extract on rhizosphere microbiome dynamics and growth enhancement in *Oryza sativa* L. RD79. *Agronomy* **2024**, *14*, 2698. [CrossRef]
7. Aslanzadeh, J. Biochemical profile-based microbial identification systems. In *Advanced Techniques in Diagnostic Microbiology*; Springer: Boston, MA, USA, 2006; pp. 84–116.
8. Brady, C.; Cleenwerck, I.; Venter, S.; Vancanneyt, M.; Swings, J.; Coutinho, T. Phylogeny and identification of *Pantoea* species associated with plants, humans and the natural environment based on multilocus sequence analysis (MLSA). *Syst. Appl. Microbiol.* **2008**, *31*, 447–460. [CrossRef]
9. Hall, T.A. BioEdit: A user-friendly biological sequence alignment editor and analysis program for Windows 95/98/NT. *Nucleic Acids Symp. Ser.* **1999**, *41*, 95–98.
10. Larkin, M.A.; Blackshields, G.; Brown, N.P.; Chenna, R.; McGettigan, P.A.; McWilliam, H.; Valentin, F.; Wallace, I.M.; Wilm, A.; Lopez, R. Clustal W and Clustal X version 2.0. *Bioinformatics* **2007**, *23*, 2947–2948. [CrossRef]
11. Kumar, S.; Stecher, G.; Li, M.; Knyaz, C.; Tamura, K. MEGA X: Molecular evolutionary genetics analysis across computing platforms. *Mol. Biol. Evol.* **2018**, *35*, 1547–1549. [CrossRef]
12. Bolger, A.M.; Lohse, M.; Usadel, B. Trimmomatic: A flexible trimmer for Illumina sequence data. *Bioinformatics* **2014**, *30*, 2114–2120. [CrossRef] [PubMed]
13. Prjibelski, A.; Antipov, D.; Meleshko, D.; Lapidus, A.; Korobeynikov, A. Using SPAdes de novo assembler. *Curr. Protoc. Bio Inform.* **2020**, *70*, e102. [CrossRef]
14. Gurevich, A.; Saveliev, V.; Vyahhi, N.; Tesler, G. QUAST: Quality assessment tool for genome assemblies. *Bioinformatics* **2013**, *29*, 1072–1075. [CrossRef] [PubMed]
15. Overbeek, R.; Olson, R.; Pusch, G.D.; Olsen, G.J.; Davis, J.J.; Disz, T.; Edwards, R.A.; Gerdes, S.; Parrello, B.; Shukla, M. The SEED and the Rapid Annotation of microbial genomes using Subsystems Technology (RAST). *Nucleic Acids Res.* **2014**, *42*, D206–D214. [CrossRef]
16. Wattam, A.R.; Davis, J.J.; Assaf, R.; Boisvert, S.; Brettin, T.; Bun, C.; Conrad, N.; Dietrich, E.M.; Disz, T.; Gabbard, J.L. Improvements to PATRIC, the all-bacterial bioinformatics database and analysis resource center. *Nucleic Acids Res.* **2017**, *45*, D535–D542. [CrossRef]
17. Grant, J.R.; Enns, E.; Marinier, E.; Mandal, A.; Herman, E.K.; Chen, C.-y.; Graham, M.; Van Domselaar, G.; Stothard, P. Proksee: In-depth characterization and visualization of bacterial genomes. *Nucleic Acids Res.* **2023**, *51*, W484–W492. [CrossRef]
18. Seemann, T. Prokka: Rapid prokaryotic genome annotation. *Bioinformatics* **2014**, *30*, 2068–2069. [CrossRef]
19. Blin, K.; Shaw, S.; Kloosterman, A.M.; Charlop-Powers, Z.; Van Wezel, G.P.; Medema, M.H.; Weber, T. antiSMASH 6.0: Improving cluster detection and comparison capabilities. *Nucleic Acids Res.* **2021**, *49*, W29–W35. [CrossRef]
20. Bortolaia, V.; Kaas, R.S.; Ruppe, E.; Roberts, M.C.; Schwarz, S.; Cattoir, V.; Philippon, A.; Allesoe, R.L.; Rebelo, A.R.; Florensa, A.F. ResFinder 4.0 for predictions of phenotypes from genotypes. *J. Antimicrob. Chemother.* **2020**, *75*, 3491–3500. [CrossRef]
21. Carattoli, A.; Zankari, E.; García-Fernández, A.; Voldby Larsen, M.; Lund, O.; Villa, L.; Møller Aarestrup, F.; Hasman, H. *In silico* detection and typing of plasmids using PlasmidFinder and plasmid multilocus sequence typing. *Antimicrob. Agents Chemother.* **2014**, *58*, 3895–3903. [CrossRef]
22. Richter, M.; Rosselló-Móra, R.; Oliver Glöckner, F.; Peplies, J. JSpeciesWS: A web server for prokaryotic species circumscription based on pairwise genome comparison. *Bioinformatics* **2016**, *32*, 929–931. [CrossRef] [PubMed]
23. Meier-Kolthoff, J.P.; Göker, M. TYGS is an automated high-throughput platform for state-of-the-art genome-based taxonomy. *Nat. Commun.* **2019**, *10*, 2182. [CrossRef]
24. Meier-Kolthoff, J.P.; Carbasse, J.S.; Peinado-Olarte, R.L.; Göker, M. TYGS and LPSN: A database tandem for fast and reliable genome-based classification and nomenclature of prokaryotes. *Nucleic Acids Res.* **2022**, *50*, D801–D807. [CrossRef]
25. Darling, A.C.; Mau, B.; Blattner, F.R.; Perna, N.T. Mauve: Multiple alignment of conserved genomic sequence with rearrangements. *Genome Res.* **2004**, *14*, 1394–1403. [CrossRef] [PubMed]

26. Yaikhan, T.; Suwannasin, S.; Singkhamanan, K.; Chusri, S.; Pomwised, R.; Wonglapsuwan, M.; Surachat, K. Genomic characterization of multidrug-resistant Enterobacteriaceae clinical isolates from Southern Thailand hospitals: Unraveling antimicrobial resistance and virulence mechanisms. *Antibiotics* **2024**, *13*, 531. [CrossRef]
27. Gong, F.; Yanofsky, C. Analysis of tryptophanase operon expression *in vitro*: Accumulation of TnaC-peptidyl-tRNA in a release factor 2-depleted S-30 extract prevents Rho factor action, simulating induction. *J. Biol. Chem.* **2002**, *277*, 17095–17100. [CrossRef]
28. Moosavian, M.; Khoshkholgh Sima, M.; Ahmad Khosravi, N.; Abbasi Montazeri, E. Detection of *OqxAB* efflux pumps, a multidrug-resistant agent in bacterial infection in patients referring to teaching hospitals in Ahvaz, Southwest of Iran. *Int. J. Microbiol.* **2021**, *2021*, 2145176. [CrossRef]
29. Bharatham, N.; Bhowmik, P.; Aoki, M.; Okada, U.; Sharma, S.; Yamashita, E.; Shanbhag, A.P.; Rajagopal, S.; Thomas, T.; Sarma, M. Structure and function relationship of *OqxB* efflux pump from *Klebsiella pneumoniae*. *Nat. Commun.* **2021**, *12*, 5400. [CrossRef]
30. Raymond, K.N.; Dertz, E.A.; Kim, S.S. Enterobactin: An archetype for microbial iron transport. *Proc. Natl. Acad. Sci. USA* **2003**, *100*, 3584–3588. [CrossRef]
31. Timofeeva, A.M.; Galyamova, M.R.; Sedykh, S.E. Bacterial siderophores: Classification, biosynthesis, perspectives of use in agriculture. *Plants* **2022**, *11*, 3065. [CrossRef]
32. Chuchert, S.; Thongmak, J.; Daungnamkaew, B.; Sudpesrot, P.; Nakorn, W.N.; Marnmad, K.; Khomphet, T. Optimizing Temperature and Drying Conditions to Break Dormancy in Thai RD79 Rice Seeds. *Indian J. Agric. Res.* **2025**, 1–9. [CrossRef]
33. Itoh, Y.; Rice, J.D.; Goller, C.; Pannuri, A.; Meisner, J.J.; Beveridge, T.J.; Preston III, J.F.; Rom, T. Roles of pgaABCD genes in synthesis, modification, and export of the *Escherichia coli* biofilm adhesin poly-β-1,6-N-acetyl-D-glucosamine. *J. Bacteriol* **2008**, *190*, 3670–3680. [CrossRef] [PubMed]
34. Lai, S.J.; Tu, I.F.; Tseng, T.S.; Tsai, Y.H.; Wu, S.H. The deficiency of poly-β-1,6-N-acetyl-glucosamine deacetylase trigger *A. baumannii* to convert to biofilm-independent colistin-tolerant cells. *Sci. Rep.* **2023**, *13*, 2800. [CrossRef]
35. Ajijah, N.; Fiodor, A.; Pandey, A.K.; Rana, A.; Pranaw, K. Plant growth-promoting bacteria (PGPB) with biofilm-forming ability: A multifaceted agent for sustainable agriculture. *Diversity* **2023**, *15*, 112. [CrossRef]
36. Rafique, M.; Naveed, M.; Mumtaz, M.Z.; Niaz, A.; Alamri, S.; Siddiqui, M.H.; Waheed, M.Q.; Ali, Z.; Naman, A.; Rehman, S.U. Unlocking the potential of biofilm-forming plant growth-promoting rhizobacteria for growth and yield enhancement in wheat (*Triticum aestivum* L.). *Sci. Rep.* **2024**, *14*, 15546. [CrossRef]
37. Lata, D.L.; Abdie, O.; Rezene, Y. IAA-producing bacteria from the rhizosphere of chickpea (*Cicer arietinum* L.): Isolation, characterization, and their effects on plant growth performance. *Heliyon* **2024**, *10*, e39702. [CrossRef]
38. Shabtai, I.A.; Wilhelm, R.C.; Schweizer, S.A.; Höschen, C.; Buckley, D.H.; Lehmann, J. Calcium promotes persistent soil organic matter by altering microbial transformation of plant litter. *Nat. Commun.* **2023**, *14*, 6609. [CrossRef]
39. Wang, C.; Luan, S. Calcium homeostasis and signaling in plant immunity. *Curr. Opin. Plant Biol.* **2024**, *77*, 102485. [CrossRef]

*agriculture*

MDPI

*Article*

# Synthetic Hydrogel Dilutes *Serratia plymuthica* Growth—Promoting Effect on *Brassica napus* L. Under Drought Conditions

**Grażyna B. Dąbrowska [1,*], Daniel Krauklis [2], Milena Kulasek [1], Magdalena Nocny [1], Marcel Antoszewski [1], Agnieszka Mierek-Adamska [1] and Beata Kaliska [2]**

1 Department of Genetics, Faculty of Biological and Veterinary Sciences, Nicolaus Copernicus University in Toruń, Lwowska 1, 87-100 Toruń, Poland; mant@doktorant.umk.pl (M.A.); mierek_adamska@umk.pl (A.M.-A.)
2 Research Centre for Cultivar Testing in Słupia Wielka, Chrząstowo 8, 89-100 Nakło nad Notecią, Poland; d.krauklis@chrzastowo.coboru.gov.pl (D.K.)
* Correspondence: browsk@umk.pl

**Abstract:** Progressive climate change increases drought frequency, severely impacting crop growth and yield. New eco-friendly alternatives to chemical protection agents and fertilisers are needed to reduce pollution and enhance plant health. This study evaluated the effects of the plant growth-promoting rhizobacteria (PGPR) and the hydrogel (potassium polyacrylate) on *Brassica napus* L. during drought conditions. After in vitro and pot experiments, *Serratia plymuthica* was selected from six PGPR candidates based on its ability to enhance plant biomass, shoot length, and number of internodes. The seed viability test, reactive oxygen species accumulation, and SOD activity showed no adverse effects of applying hydrogel to canola seeds. Moreover, the hydrogel did not show toxicity towards tested bacterial strains. Field trials during the drought demonstrated that inoculation with *S. plymuthica* significantly increased the number of siliques (16.48%) and yield (19.93%), compared to controls. Plants grown from inoculated seeds were also taller (3.28%) and had more branches (39.99%). We also analysed seedling emergence, number of internodes, thousand seeds' weight, and seed moisture. The hydrogel applied to the soil delayed seedling emergence and dampened the growth-promoting effect of *S. plymuthica*, resulting in reduced yield. Compared with plants inoculated with *S. plymuthica*, there was a decrease in the yield of plants treated solely with hydrogel and in plants treated with hydrogel and the bacterium of 23.61% and 27.79%, respectively. Our results indicate that *S. plymuthica* has a high potential for use in commercial canola farming, especially as an ingredient added to seed coatings.

**Keywords:** bioinoculants; canola; drought stress; potassium polyacrylate; PGPM; rhizobacteria; ROS

## 1. Introduction

The latest IPCC (The Intergovernmental Panel on Climate Change) report highlights that climate change is expected to significantly increase droughts' frequency, duration, and intensity in many regions. It emphasizes that rising temperatures, changes in precipitation patterns, and increased evaporation rates contribute to more severe drought conditions [1]. Drought affects various essential processes in plants, including respiration [2], photosynthesis [3], stomatal conductance and movement [4], and disturbs plant water use efficiency [5] and nutrient acquisition [6,7]. In this way, drought affects plant growth and development,

 https://doi.org/10.3390/agriculture15020142

significantly influencing plant productivity and consequently leading to yield loss [3]. Estimated yield loss ranges from 30 to 90%, depending on crop species and drought severity [8]. Furthermore, with the predicted increase in the global population, drought presents a significant threat to food security, affecting both the quantity and quality of food [9].

Canola (*Brassica napus* L. var. *napus*) is the second most widely cultivated oilseed crop worldwide, with 2020/2021 production reaching 73.6% Mt [10]. Increasing canola production is linked to its favourable fatty acids composition, well-balanced amino-acids content, and many applications in the industry, i.e., biofuels, paints, polyurethanes, adhesives, and coatings [11]. Canola is sensitive to water shortages at all stages of vegetative growth and seed development and maturing [12], with germination and early seedling development being the most sensitive stages, resulting in decreased biomass and survival rates [13]. Water deficit during flowering and seed maturation can result in altered seed content composition, i.e., an increase in protein content with a simultaneous decrease in phenolics and fatty acids, and an overall decreased number of siliques [12].

One proposed solution for enhancing plant health and yield in agriculture is seed priming [14,15] and/or plant growth-promoting microorganisms (PGPM) [16–18]. Bioinoculants comprised of living or dormant organisms positively influence plant growth and development through enhancement of the germination parameters [19,20], increased elongation of the shoot and root [21–23], increased biomass production and yield [24,25], acceleration of the flowering [26,27], and enhanced photosynthesis efficiency [4,28]. Furthermore, PGPM can ameliorate stress effects and improve plant stress tolerance to both biotic [29,30] and abiotic stresses, including heavy metal pollution [31,32], salinity [20,33], and drought [4,25,34]. Inoculation of wheat (*Triticum aestivum* L.) with *Burkholderia phytofirmans* PsJN ameliorated the effect of the drought stress, improved photosynthetic efficiency, and improved yield by 18–21%. Moreover, this procedure improved water use efficiency and ionic balance and increased nitrogen, phosphorus, potassium, and protein content in the grain [4]. In a greenhouse experiment, inoculation of maize (*Zea mays* L.) seeds and maize rhizosphere with *Azospirillum* sp. resulted in increases in free amino-acid (54% and 45%, respectively) and soluble sugar content (63% and 31%, respectively) under drought stress relative to the control plants (non-inoculated) [34]. This highlights the dramatic impact of the inoculation method on the outcome. In field conditions, *B. napus* L. seeds inoculated with *Trichoderma viride* VII showed an increase in yield of 16.7% during drought compared to non-inoculated plants. Furthermore, *T. viride* VII positively influenced the height and branching of the plants, the number of siliques, and the weight of the thousand seeds (TSW) [25]. A recent meta-analysis on the PGPR (plant growth-promoting rhizobacteria) showed that rhizobacteria inoculation of well-watered plants increases root and shoot biomass by 35% and 28%, respectively, with an average yield increase of 19%. Interestingly, those numbers are higher for plants under drought stress conditions, with a 43% and 45% increase in root and shoot biomass, respectively, and a 40% increase in yield [35]. A similar observation was noted by Zhao et al. [36], where authors showed that PGPR more efficiently increases plant biomass, improves photosynthesis rate, and more efficiently ameliorates oxidative stress in plants subjected to water deficit than in control plants [36].

Another proposed approach to mitigate drought effects on arable lands is to supplement the soil with hydrogel: hydrophilic gels that can absorb water via hydrogen bonds with gel hydrophilic functional groups (-OH, -COOH, and -$NH_2$). Polymers that can serve as hydrogel bases include collagen, gelatine, alginate, and agarose [37]. Hydrogels have a unique capacity to hold large volumes of water in the polymer network and, additionally, their porous structure slows evaporation [4]. Agricultural hydrogels can increase the water-holding capacity of soil [37,38] and, thus, contribute to water saving, improve soil moisture, prevent the leaching of nutrients from fertilisers, and reduce soil erosion [37]. Zhu et al. [39]

demonstrated that applying hydrogel derived from bean curd and soybean milk to soil increased the yield of leafy vegetables by over 60% and 35% in choy sum (*Brassica rapa* var. *parachinensis* (L.H. Bailey) Hanelt) and pak choi (*Brassica rapa* subsp. *chinensis* (L.) Hanelt), respectively, compared to control plants grown without hydrogel. Furthermore, these plants exhibited greater leaf area and improved water use efficiency under field conditions. Interestingly, the effects of the hydrogel supplementation depend greatly on the soil type they are used on. Makowska et al. [40] showed that applying the hydrogel to black soil and sandy soil increased strawberry (*Fragaria* × *xananassa* Duchesne) water content under field conditions, with twice higher values for black soil than for sandy soil. Moreover, the hydrogel application at a concentration of 3 g/dm$^3$ improved stomatal conductance and photosynthesis efficiency. In comparison, applying the hydrogel at a higher concentration (6 g/dm$^3$) seemed to dilute its positive effect on tested parameters. Nevertheless, the results were still higher than in control plants not treated with the hydrogel.

Although, recently, there has been a spike in the number of studies on the use of PGPR and hydrogel for improving crop performance under various conditions, there is a clear gap regarding data on how PGPR and hydrogel perform under field conditions. This study aimed to find the best combination of PGPM–hydrogel that would protect *B. napus* from drought and improve the crop performance of canola in a field experiment. We hypothesized that the simultaneous application of hydrogel and rhizobacteria in canola cultivation would synergistically affect growth promotion and yield increase. First, we evaluated the growth-promoting and drought-stress-ameliorating capacity of six bacterial strains in a pot experiment and selected *S. plymuthica* for further testing in the field experiment. We have also tested the hydrogel's effect on the bacteria and the potential toxic effect on canola seeds.

## 2. Materials and Methods

### 2.1. Biological Material

Seeds of spring canola (*Brassica napus* L.), variety Lumen, were purchased from Rapool (Wągrowiec, Poland). Unless otherwise stated, the seeds were surface-sterilized according to the method presented by Mierek-Adamska et al. [41] in a mixture of 96% ethanol and 30% $H_2O_2$ (1:1, *v:v*) for 5 min and then rinsed at least 10 times with sterile water. Surface-sterilized seeds were placed on sterile Petri dishes containing filter paper discs soaked in 3 mL of sterile water. The dishes were transferred to a growth chamber at 25 °C for a 12/12 h light/dark photoperiod for three days.

Bacteria used in this study come from the collection of the Department of Genetics at the Nicolaus Copernicus University (NCU) in Toruń: *Bacillus* sp., *Bacillus megaterium*, *Bacillus polymyxa*, and *Paenibacillus peoriae* and the Department of Microbiology (NCU, Toruń): *Serratia plymuthica* and *Serratia liquefaciens*. Bacterial strains were stored at −80 °C in glycerol stocks (80%).

### 2.2. Evaluation of Bacterial Growth in the Presence of the Hydrogel

Bacterial strains from glycerol stocks were inoculated onto solid R2A medium. Petri dishes with inoculated bacteria were incubated at 25 °C for 2–5 days, depending on the growth rate of the bacterial strain. Then, liquid cultures were established by inoculating 20 mL of R2A liquid medium, which were maintained for 2–5 days at 25 °C on a laboratory shaker with a thermostat. Then, the optical density of bacterial cultures was measured and diluted to the optical density $OD_{600}$ = 0.2. To test the effect of the hydrogel on bacteria, 10 mg of hydrogel (potassium polyacrylate (AgroManager, Osielsko, Poland)) and 100 μL of a suspension of a single bacterial strain were added to 5 mL of R2A medium. The tubes were incubated at 25 °C using a shaking platform for 5 days. The control consisted of bacterial liquid cultures without hydrogel. Liquid cultures of individual bacterial strains were centrifuged for 10 min at 12,000 rpm using a minicentrifuge (BioRad, Hercules, CA, USA). The supernatant was carefully removed, and the bacterial pellet was weighed on a laboratory scale to assess the fresh biomass, then lyophilised using the CentriVap Cold Trap system (Labconco, Kansas City, MO, USA) and weighed to determine the dry biomass. Three replicates were performed for each bacterial strain.

### 2.3. Pot Experiment

Canola seeds were placed in pots (5 seeds per pot) with 1 g of hydrogel per 1 L of soil (500 mL sand mixed with 500 mL of garden soil). The sterile seeds were inoculated (soaked in bacterial suspension in $dH_2O$ at $5.0 \times 10^7$ CFU/$cm^3$) before sowing or not (control; soaked in sterile $dH_2O$), according to the protocol described previously [33]. The plants were watered ad libitum for five weeks. Then, watering was stopped to induce physiological drought stress. This stress lasted until drought symptoms appeared (leaf drying and lack of turgor), which were visible after two weeks of water deprivation. At that time, the above-ground part of the plants was collected, and the height of the plants, the number of internodes, and their dry biomass were measured. The experiment was performed in 3 biological replicates.

### 2.4. ROS Localisation

Hydrogen peroxide was localised according to Daudi et al. [42]. Seedlings were infiltrated in the dark with shaking for 3 h in a 0.1% 3,3′-diaminobenzidine (DAB) solution in 10 mM phosphate buffer containing 0.05% TWEEN-20. After infiltration, the samples were rinsed twice with distilled water and decolourised in a mixture of ethanol, glycerol, and acetic acid in a 3:1:1 ratio ($v/v/v$). For $O_2^-$ radical localisation, nitro blue tetrazolium chloride (NBT) staining method was used [30]. Canola seedlings were infiltrated in the dark for 2 h with a 0.1% nitro blue tetrazolium chloride (NBT) and 10 mM sodium azide in 50 mM phosphate buffer (pH 7.5). After infiltration, the samples were rinsed twice with distilled water and decolourised in a mixture of ethanol, glycerol, and acetic acid in a 3:1:1 ratio ($v/v/v$). Negative controls were incubated with solutions lacking DAB or NBT. The experiment was performed in triplicate.

Numeric data for semi-quantitative analysis were obtained using ICY software (https://icy.bioimageanalysis.org, accessed on 15 November 2024) and expressed as a summary of total red (DAB) or blue colour (NBT) pixel intensity divided by the sum of red, blue, and green pixel intensities.

### 2.5. SOD Activity

Superoxide dismutase (SOD) activity was measured via the spectrophotometric method [31]. To isolate total proteins, 50 mg of plant tissue (3-days seedlings) was ground in liquid nitrogen and suspended in 1 mL of cold protein isolation buffer (100 mM tricine, 3 mM $MgSO_3$, 3 mM EGTA, 1 mM DTT, pH 7.5 adjusted with 1 M Tris). Then, it was incubated on ice for 15 min and centrifuged at 13,000× $g$ for 20 min at 4 °C. The protein concentration in the resulting extract was measured using the Pierce™ Coomassie (Bradford) Protein Assay Kit and multidetection microplate reader (Synergy model HT; BioTek, Winooski, VT, USA).

For superoxide dismutase (SOD) activity measurements, the working solution consisted of a mixture of 0.1 M phosphate buffer pH 7.5, 2.4 μM riboflavin, 840 μM NBT, 150 mM methionine, and 12 mM $Na_2EDTA$ in a ratio of 8:1:1:1:1 ($v/v/v/v/v$). Two microtiter plates were prepared, with 5 μL of the protein extract added in triplicate for technical and two biological replicates. Subsequently, 150 μL of the working solution was added to each well. One plate was kept in the dark as a blank, while the other was exposed to warm white LED light at 400 μE for 15 min. Absorbance at 560 nm was measured using a multidetection microplate reader (Synergy model HT; BioTek, Winooski, VT, USA).

### 2.6. Tetrazolium (TZ) Test

The seeds were incubated in a 1% solution of 2,3,5-triphenyl tetrazolium chloride (TTC) in PBS buffer (pH 7.4) for 24 h at 30 °C in the dark. After incubation, the samples were rinsed three times with distilled water and decolorised in a mixture of lactic acid, phenol, glycerine, and water in a 1:1:2:1 ratio (*v*/*v*/*v*/*v*). The seed shells were removed to visualise the embryos. Negative controls consisted of seeds boiled for 1 h at 95 °C.

Numeric data for semi-quantitative analysis were obtained using ICY software and expressed as a total red pixel intensity divided by the sum of red, blue, and green intensities.

### 2.7. Field Experiment

The field experiment was conducted, as presented in our previous study [25], in Chrząstowo (53°09′52″ N 17°35′02″ E), on a site with medium clay, brown soil pH 6.4. The experiment was established on 7 April 2022, and the crop was harvested on 12 August 2022. The content of bioavailable minerals was analysed at the Chemical-Agricultural District Station in Bydgoszcz, Poland, and it amounted to K—183 mg/kg$^{-1}$, Mg—47 mg/kg$^{-1}$, and P—95 mg/kg$^{-1}$. Non-sterile canola seeds were sprayed with sterile $H_2O$ (control) or with bacterial suspension in $dH_2O$ at 5.0 × $10^7$ CFU/cm$^3$ (*S. plymuthica*); seeds were constantly mixed to ensure even spread of the inoculum, then dried before sowing. The size of the experimental field was 16.5 m$^2$. Before sowing, the hydrogel was put into the soil (4 cm deep) with the seeder, according to the manufacturer's instructions. The hydrogel was applied in 0.5 m wide soil strips, into which inoculated and non-inoculated seeds were mechanically sown (2 cm deep; density of 100 plants per m$^2$) using a sowing machine. The experimental plots were randomised, and the experiment was carried out simultaneously in three replicates on different plots next to each other. Fertilisation was carried out as described in the previous article [25], with half-dose before sowing, and the other half was used afterward as a top dressing, with total doses of the pure ingredients of 30 $NH_4NO_3$ kg/ha; 48 $P_2O_5$ kg/ha; 34 $(NH_4)_2SO_4$ kg/ha; and 80 $K_2O$ kg/ha. The insecticides Decis Mega 50EW (Bayer; Leverkusen, Germany) and Boravi 50 WG (Gowan; Yuma, AZ, USA) were used to protect the plants against silique pests. After harvesting, the following canola parameters were evaluated on 10 randomly selected plants from each of three plots: height, number of branches, and siliques. Moreover, the seed yield was determined at 9% humidity of seeds, and thousand seed weight (TSW) was measured. The emergence of seedlings was evaluated according to the BBCH scale.

A weather station installed at the field site registered the precipitation and temperature. Daily precipitation was measured using a Hellmann pluviometer (capacity 200 cm$^2$; Herter, Poland) stationed 1 m above the ground. The daily average temperature was calculated based on readings at 8:00, 14:00, and 20:00 with the formula:

$$\text{Average daily temperature, TD} = (T_{min} + T_{max} + T_1 + T_2)/4$$

where $T_{min}$ is the lowest temperature recorded, $T_{max}$ is the highest temperature recorded, $T_1$ is the temperature measured at 8:00, and $T_2$ is the temperature measured at 20:00.

During the growing season, there were unfavourable environmental conditions, i.e., ground frost in April (down to −4.5 °C) and May (down to −6.2 °C), and high air temperatures (in June up to 34.6 °C and in July up to 35.7 °C). Precipitation and average temperature during the growing season are shown in Figure 1.

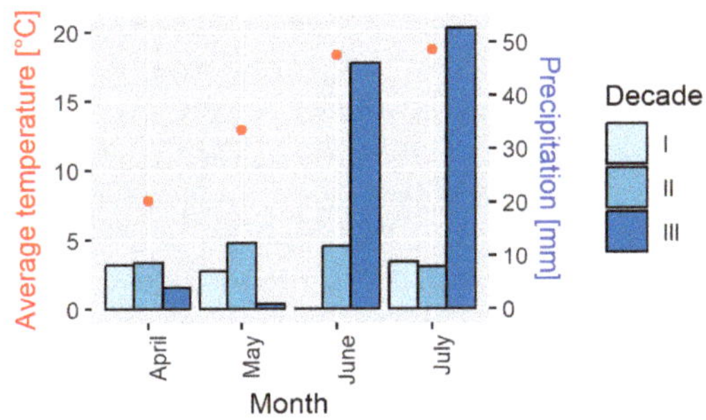


**Figure 1.** The precipitation and average temperatures during the canola growing season. Decade denotes 10 consecutive days during a month.

*2.8. Statistical Analysis*

Data were analysed with a one-way analysis of variance (ANOVA) with a Tukey post hoc test in R [43] using the following packages: 'ggplot2', 'dplyr', 'tidyr', 'FSA', 'multcomp', 'multcompView', and 'plyr'.

## 3. Results

*3.1. Bacterial Growth in the Presence of the Hydrogel*

The fresh biomass of *S. plymuthica* was over 3.7% higher in the presence of the hydrogel than in the control conditions (without hydrogel). However, we did not observe a similar significant improvement in the dry biomass (Figure 2). The growth of *B. megaterium* and *B. polymyxa* showed a similar trend to *S. plymuthica*, with a substantial increase in the fresh biomass by 3.83% and 5.02%, respectively, in the presence of the hydrogel. The growth of *S. liquefaciens*, *P. perioae*, and *Bacillus* sp. showed no significant changes in these conditions.

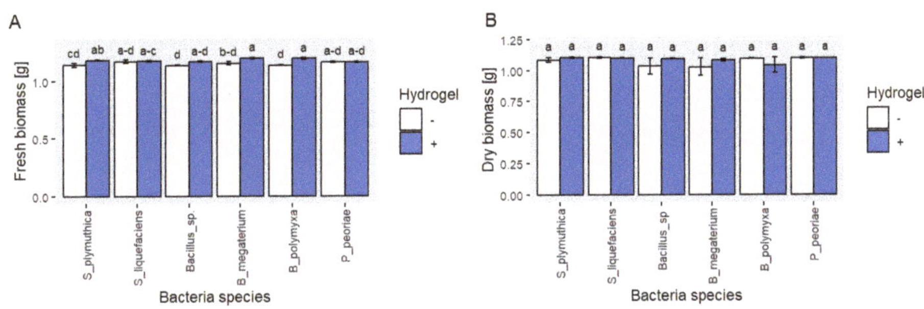


**Figure 2.** (**A**) The fresh and (**B**) dry biomass of bacteria strains (*B. megaterium*, *B. polymyxa*, *Bacillus* sp., *P. peoriae*, *S. liquefaciens*, and *S. plymuthica*) grown in the liquid R2A medium with (blue bars) or without (control, white bars) the hydrogel. Bars represent means with standard error. Statistical analysis was performed using one-way ANOVA followed by Tukey's post hoc test. Different letters indicate statistically significant differences ($p > 0.05$).

*3.2. Impact of PGPR and Hydrogels on Canola Growth*

The application of the hydrogel seemed to dilute the growth-promoting activity of some strains (*S. plymuthica*, *S. liquefaciens*, and *P. peoriae*) and enhance this effect in the case of bacteria belonging to the Bacillus genus (*Bacillus* sp., *B. megaterium*, and *B. polymyxa*) (Figure 3). The highest results for the shoot length were noted in *B. polymyxa* with added hydrogel and *S. plymuthica* without the hydrogel, with 24% and 20% increase, respectively, when compared to the control, and around 34–39% when compared to non-inoculated plants grown in the presence of the hydrogel. The lowest scores were noted for *P. peoriae* with hydrogel, where the overall height was lower by over 38% compared to the control and by 23% compared to the hydrogel (non-inoculated plants). Furthermore, applying the hydrogel to the soil of pots with non-inoculated plants reduced plant height by over 12% compared to control plants. The highest number of internodes was noted for plants inoculated with *S. plymuthica* and *B. megaterium*, both with and without the addition of the hydrogel, with 2–3 more internodes in comparison to the control plants (Figure 3). The dry biomass was highest in plants inoculated with *S. plymuthica*, irrespective of the addition of hydrogel, with an increase in the range of 36–54% in comparison to control plants (non-inoculated, without hydrogel) and 46–65% compared to non-inoculated plants grown in the presence of hydrogel (Figure 3).

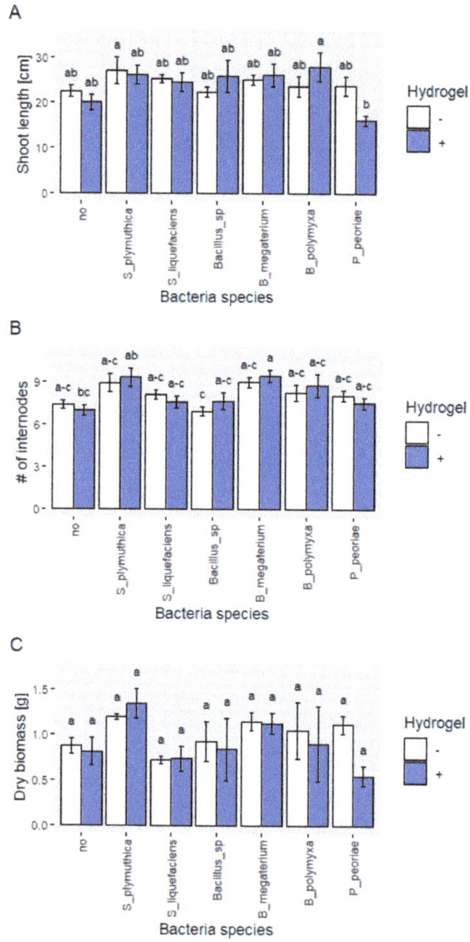


**Figure 3.** Biometric parameters of *B. napus* cultivated in bacterial strains' absence in the pot experiment conditions with (blue bars) or without (control, white bars) the hydrogel. (**A**) shoot length of canola; (**B**) number of internodes of canola; and (**C**) dry biomass of canola. Bars represent means (n = 15 for shoot length and number of internodes and n = 3 for dry biomass) with standard error. Statistical analysis was performed using one-way ANOVA followed by Tukey's post hoc test. Different letters indicate statistically significant differences ($p > 0.05$).

### 3.3. Antioxidant Balance

Canola seedlings grown in the presence of hydrogel did not exhibit a higher accumulation of reactive oxygen species: $H_2O_2$ (Figure 4A–D,I) or $O_2^-$ (Figure 4E–H,J). Similarly, seeds and seedlings with and without the hydrogel did not display differences in superoxide dismutase (SOD) activity (Figure 4K). These results indicate that the hydrogel does not cause stress in canola seedlings.

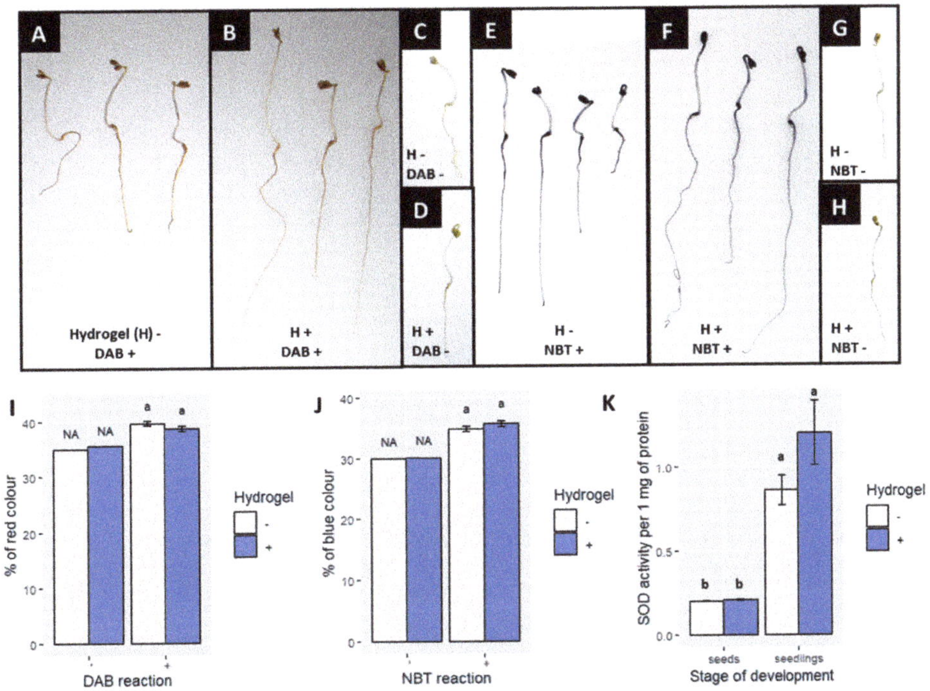


**Figure 4.** Reactive oxygen species generation and scavenging in canola seedlings under control conditions (H$^-$) and hydrogel treatment (H+). (**A–D**) DAB staining, where the brown colour indicates $H_2O_2$ accumulation. (**E–H**) NBT staining, with navy blue indicating $O_2^-$ accumulation. (**A**,**E**) control plants; (**B**,**F**) plants growing in the hydrogel; and (**C**,**D**,**G**,**H**) negative controls for the staining procedures. Bar plots (**I**,**J**), results of semi-quantitative measurement of $H_2O_2$ and $O_2^-$ accumulation. (**K**) the activity of SOD in seeds and seedlings. Bars represent means (n = 15 for the shoot length and number of internodes and n = 3 for dry biomass) with standard error. Statistical analysis was performed using one-way ANOVA followed by Tukey's post hoc test. Different letters indicate statistically significant differences ($p > 0.05$).

### 3.4. Seed Vitality

Tetrazolium (TZ) staining showed no decrease in the vitality of the canola seeds germinated in control conditions or in the presence of the hydrogel (Figure 5), which is similar to the analyses concerning antioxidant balance.

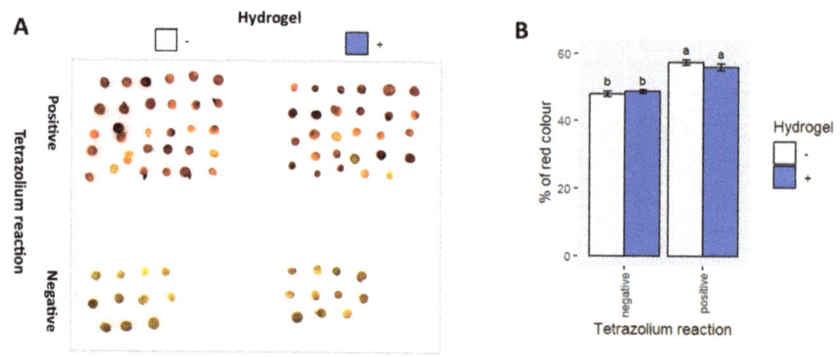


**Figure 5.** The tetrazolium (TZ) test assesses the germinability of canola under control conditions and with hydrogel treatment. The negative control consists of seeds incubated for 1 h at 95 °C. (**A**) Photographs of stained seeds. (**B**) Semi-quantitative evaluation of the TZ reaction, calculated as the percentage of red pixels. Statistical analysis was performed using one-way ANOVA followed by Tukey's post hoc test. Different letters indicate statistically significant differences ($p > 0.05$).

### 3.5. *Canola Growth in the Presence of Hydrogel and S. plymuthica in the Field Experiment*

The hydrogel negatively affected seedling emergence in canola (Figure 6). Moreover, plants grown in the presence of hydrogel but not inoculated with *S. plymuthica* were, on average, smaller by 4–8 cm than control plants and inoculated plants. Yield and TSW were lowest in plants grown with the hydrogel. Canola inoculated with *S. plymuthica* showed a significant increase in the number of siliques (16.48%), yield (19.93%), and TSW (2%) when compared to control plants. Applying hydrogel to plants inoculated with *S. plymuthica* weakened the bacterium's growth-promoting effect, resulting in decreased yield (Figure 6).

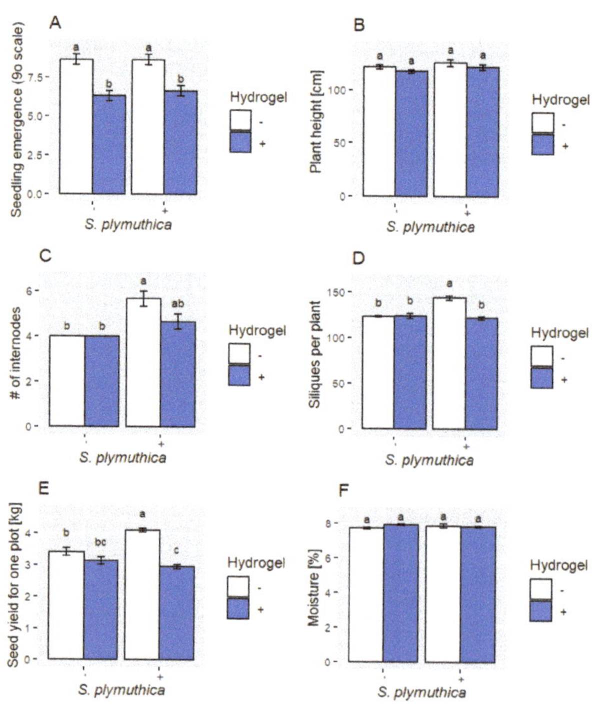


**Figure 6.** *Cont.*

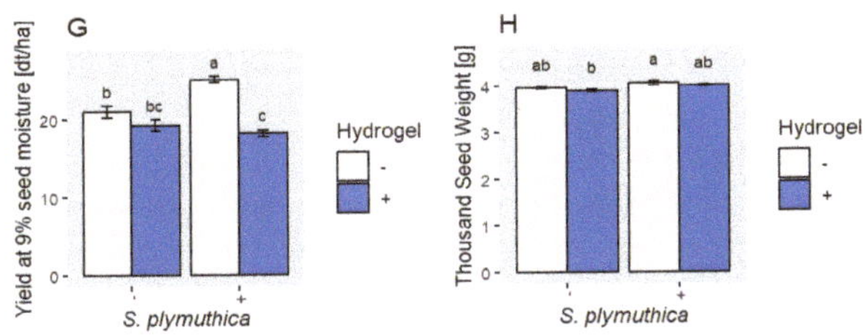


**Figure 6.** Biometric evaluation of *B. napus* L. plants grown from seeds inoculated with *S. plymuthica*, soil mixed with hydrogel, and soil mixed with both hydrogel and *S. plymuthica* inoculated seeds under field conditions. (**A**) seedling emergence; (**B**) plant height; (**C**) number of internodes; (**D**) siliques per plant; (**E**) seed yield for one plot; (**F**) moisture content of seeds; (**G**) yield at 9% of moisture; and (**H**) thousand seeds weight. Statistical analysis was performed using one-way ANOVA followed by Tukey's post hoc test. Different letters indicate statistically significant differences ($p > 0.05$).

## 4. Discussion

We hypothesised that combining two agrotechnical treatments, i.e., simultaneous use of the hydrogels and inoculation with PGPM can synergistically enhance canola growth and development in field conditions. Applying bioinoculants in field conditions yields mixed effects, mainly due to the varying survival levels of microbes introduced to the soil. Factors influencing the successful establishment of the introduced PGPM strain include environmental conditions (i.e., temperature, rainfall, and soil type) [17,44] and the ability to compete with the native soil microbiome [17,45]. Therefore, there is an urgent need to evaluate further PGPM strains in the laboratory environment (i.e., pot experiments under greenhouse conditions) and field experiments. To verify our hypothesis, we evaluated the growth of bacterial isolates in hydrogel-enriched media to confirm that it was not toxic to the bacteria. Our findings indicated that while hydrogel addition to the growth medium improved the fresh biomass of the tested bacteria, it did not significantly affect the dry biomass. This discrepancy may suggest that the hydrogel particles adhered to bacterial cells, altering biomass measurements, as the dry weight of bacteria grown with the hydrogel was similar to that of the control. A seed viability test (TTC) was conducted to evaluate hydrogel's potentially harmful or toxic activity toward the seeds. TTC and ROS accumulation analysis showed no adverse effect of the hydrogel on the canola seeds. Moreover, SOD activity remained unchanged in seeds and seedlings, indicating that no stress response occurred in the plants. These results suggest that the hydrogel is safe for seed coatings and application to the soil and is non-toxic to the rhizobacteria. Mathes et al. [46] showed that polyacrylic acid (PAA), and PAA with the addition of lignocellulose fibres, positively affected the ingression of the tested rhizobacteria to the wheat root system. Moreover, the authors showed that applying the tested hydrogels increased the number of taxa associated with phosphorus solubilisation in the field experiment. Interestingly, the microbiome of the wheat rhizosphere had a differential composition compared to the bulk soil and control plants rhizosphere (i.e., not treated with hydrogel).

The pot experiment demonstrated that all tested bacteria enhanced canola growth under simulated drought conditions, enabling us to select *S. plymuthica* as the most promising candidate for further field studies. Our previous studies (pot experiment) have shown that the selected *S. plymuthica* strain exhibits proteolytic, cellulolytic, and ligninolytic activity and can promote canola growth in compost soil and agricultural soil contaminated with plastics [47,48]. These bacteria can form biofilm on the surface of plastic polymers [49], indicating strong rhizocompetence. Proença et al. [50] showed that *S. plymuthica* can solubilise phosphates and zinc; produce siderophores, indole-3-acetic acid (IAA), and ACC deaminase; and exhibit chitinolytic and nematicidal activity. Moreover, treating canola with

PGPM can reduce the occurrence of pathogenic microbes. Abuamsha et al. [51] reported that coating canola seeds with *S. plymuthica* reduces the infection rate of *Leptosphaeria maculans*. *S. plymuthica* also demonstrates antagonistic activity towards other fungal pathogens, including *Verticillium bilit*, *Rhizoctonia solani*, and *Sclerotinia sclerotiorum* [52]. Applying *S. plymuthica* positively impacted the canola growth under drought conditions, increasing plant height, number of siliques, branching, and TSW. Applying *S. plymuthica* alone resulted in the most favourable results for canola growth and yield, which was consistent with the previous research on the effects of *Serratia* spp. on the growth of various plants, including wheat (*Triticum aestivum* L.) [53], maize [54,55], chilli pepper (*Capsicum annum* L.) [56], black pepper (*Piper nigrum* L.) [57], rice (*Oryza sativa* L.) [58,59], and quinoa (*Chenopodium quinoa* Willd.) [60]. The observed enhanced silique formation and TSW suggest that *S. plymuthica* may have increased nutrient uptake and/or availability via Zn and P solubilisation, as well as chelation of Fe through siderophores exudation. Increased height and branching in inoculated canola are likely associated with the ability of *S. plymuthica* to produce IAA [50]. *S. plymuthica* has been reported to enhance plant growth parameters in ornamental plants under water stress, boosting flowering and shoot dry biomass in petunia (*Petunia* × *hybrida* (Sweet) D. Don ex W. H. Baxter), impatiens (*Impatiens walleriana* Hook.f.), and pansy (*Viola* × *wittrockiana* Gams ex Nauenb. & Buttler) [52]. Similar results were found in wheat grown under severe water deficit, where *Serratia* sp. 1–9 significantly increased wheat biomass and root area [61]. The interesting point is that there is still a need to evaluate specific bacterial species' interaction with particular crops. Pan et al. [62] showed that *Serratia proteamaculans* and *Serratia liquefaciens* had no apparent effect on the growth and nodulation of field-grown soybean (*Glycine max* L. Merr), whereas *S. liquefaciens* significantly improved the biomass of radish (*Raphanus sativus* L.) under heavy metal stress [63].

The use of the hydrogel yielded less desirable results than expected. Canola plants showed stunted seedling emergence, reduced height, and slightly lower TSW. These results indicate that hydrogels may have created overly moist conditions in the root zone, reducing oxygen availability and impairing root development, affecting the seedling's emergence [64]. Although hydrogels are generally reported to increase soil porosity and oxygen availability [65], our results highlight an urgent need for further research on using agricultural hydrogels in field conditions. The simultaneous use of hydrogel and *S. plymuthica* yielded results similar to those obtained for the variant with hydrogel alone, indicating that the combination of soil amendments (e.g., agricultural hydrogels) can interfere with or diminish the growth-promoting effects of rhizobacteria. The specific properties of the hydrogel used may have hindered the expected synergy between hydrogel and PGPM. Its high water-retention capacity could have reduced oxygen availability, limiting microbial colonisation and activity in the rhizosphere. Furthermore, commercially available synthetic hydrogels often contain acrylamide, acrylic acid, or potassium acrylate, which possess a negative charge [66], and the inclusion of NaOH or KOH as gelinatisation initiators and solving agents may lead to the accumulation of Na and K ions in the rhizosphere, leading to salinisation [67,68]. Some reports suggest that excess Na may impair the uptake of Ca and lead to root rot and reduced biomass accumulation [66,69]. Moreover, high concentrations of solutes in the hydrogel may lead to osmotic stress and impair the water uptake by the root system [66]. While simultaneous application of the hydrogel and *S. plymuthica* improved canola branching, the yield and seedling emergence were lower when compared to control plants, which may indicate that hydrogel somewhat impaired the ability of the bacteria to colonise the rhizosphere or limited the oxygen availability to the roots, effectively negating the plant growth promoting abilities of the bacterium [64]. While some studies show the variable effectiveness of hydrogels depending on environmental factors and plant species, comprehensive research on hydrogel–soil–plant interactions in field conditions is still needed. Hydrogels not only increase the water-holding capacity of the soil but also limit the leaching of nutrients from the soil [65,70]. As mentioned before, the soil type greatly impacts the efficiency of the applied hydrogel. Womack et al. [71] showed an increase in soil porosity of 4, 10, and 19% in clay, sandy, and sandy loam soils, respectively. Moreover, the authors reported that naturally derived hydrogels had a more favourable impact on the soil structure than commercial synthetic ones [71]. Shen et al. [72] assessed the survival of *Serratia plymuthica* A21-4 in pot and field experiments. Adding 2% hydrogel to the soil in the pot experiment increased the bacterial colonisation of both the soil and the rhizosphere of pepper (*Capsicum* L.). A 100-fold higher number of bacteria was demonstrated in the rhizosphere of 40-day-old pepper plants in the soil treated with the hydrogel compared to the control. Pepper plants treated with the hydrogel in the pot experiment also showed a higher seed germination rate and better growth, as well as reduced *Phytophthora* blight.

Although hydrogels are known to increase soil porosity and water retention, these same properties can create overly moist conditions in certain soils, leading to reduced oxygen availability and stunted root development [64]. This paradox highlights the need to explore further our understanding of hydrogel application across varying environmental and soil conditions, particularly field conditions, and in the context of the rhizosphere microbiome.

The use of PGPR, such as *S. plymuthica*, may present a cost-effective alternative to conventional fertilisers since prices of fertilisers are on a continuous rise [73,74]. The potential for increased yield and reduced dependency on chemical fertilisers could offer significant economic benefits to farmers, particularly in regions facing water scarcity. Furthermore, reduced use of synthetic chemicals and environmental and regulatory benefits frame the PGPR as a sustainable long-term solution. However, the cost of hydrogel production and application remains a concern, as its benefits may vary based on soil type and environmental conditions [68,71], requiring careful cost–benefit analysis before widespread adoption.

## 5. Conclusions

Our study demonstrates *S. plymuthica's* ability to improve the crop performance of the canola under drought stress, which highlights its potential for use in commercial farming. Moreover, the successful use of the PGPR in increasing yield with simultaneously reduced fertilization, as required by the European Union, confirms its suitability for sustainable farming practices. The hydrogel was found ineffective in improving canola growth under the field conditions and interfered with seedling emergence, which points to a need for further evaluation of commercially available synthetic hydrogels. Moreover, the hydrogel diluted the growth-promoting effect observed for *S. plymuthica*, which implies that to achieve a synergistic effect in improving crop performance, there is an urgent need for further studies to understand plant–microbe–hydrogel interactions under varying field conditions. An interesting research point would be to observe the long-term effect of the hydrogel application on soil health and the microbiome.

**Author Contributions:** Conceptualization, G.B.D.; methodology, G.B.D., M.K., M.N., B.K. and D.K.; resources, G.B.D. and M.A.; software, M.A. and M.K.; validation, M.N., D.K., M.K. and A.M.-A.; formal analysis, M.K., M.N., D.K. and M.A.; investigation, D.K., M.K., M.N. and A.M.-A.; data curation, M.K., M.N., M.A. and D.K.; writing—original draft preparation, G.B.D., M.A. and M.K.; writing—review and editing G.B.D., M.A., M.K. and A.M.-A.; visualization, M.K., M.A. and D.K.; supervision, G.B.D., A.M.-A., M.K. and B.K.; project administration, G.B.D.; funding acquisition, G.B.D. and B.K. All authors have read and agreed to the published version of the manuscript.

**Funding:** This research was funded by The Marshal's Office of the Kuyavian-Pomeranian Voivodeship, grant number 690 40001.0000011 for G.B.D.

**Institutional Review Board Statement:** Not applicable.

**Data Availability Statement:** Relevant data applicable to this research are included in the paper and are also available upon request from the corresponding author.

**Acknowledgments:** The authors would like to thank Katarzyna Hrynkiewicz from the Department of Microbiology of Nicolaus Copernicus University of Toruń for providing the bacterial strains for research.

**Conflicts of Interest:** The authors declare no conflicts of interest.

## References

1. Lee, H.; Romero, J. *Climate Change 2023: Synthesis Report. Contribution of Working Groups I, II and III to the Sixth Assessment Report of the Intergovernmental Panel on Climate Change*; IPCC: Geneva, Switzerland, 2023.
2. Metcalfe, D.B.; Meir, P.; Aragão, L.E.O.C.; Lobo-do-Vale, R.; Galbraith, D.; Fisher, R.A.; Chaves, M.M.; Maroco, J.P.; da Costa, A.C.L.; de Almeida, S.S.; et al. Shifts in plant respiration and carbon use efficiency at a large-scale drought experiment in the eastern Amazon. *New Phytol.* **2010**, *187*, 608–621. [CrossRef] [PubMed]
3. Shah, N.H.; Paulsen, G.M. Interaction of drought and high temperature on photosynthesis and grain-filling of wheat. *Plant Soil* **2003**, *257*, 219–226. [CrossRef]
4. Naveed, M.; Hussain, M.B.; Zahir, Z.A.; Mitter, B.; Sessitsch, A. Drought stress amelioration in wheat through inoculation with *Burkholderia phytofirmans* strain PsJN. *Plant Growth Regul.* **2014**, *73*, 121–131. [CrossRef]
5. Ghannoum, O.; Von Caemmerer, S.; Conroy, J.P. The effect of drought on plant water use efficiency of nine NAD–ME and nine NADP–ME Australian C4 grasses. *Funct. Plant Biol.* **2002**, *29*, 1337–1348. [CrossRef]
6. Bista, D.R.; Heckathorn, S.A.; Jayawardena, D.M.; Mishra, S.; Boldt, J.K. Effects of drought on nutrient uptake and the levels of nutrient-uptake proteins in roots of drought-sensitive and -tolerant grasses. *Plants* **2018**, *7*, 28. [CrossRef]
7. Haile, G.G.; Tang, Q.; Li, W.; Liu, X.; Zhang, X. Drought: Progress in broadening its understanding. *Wiley Interdiscip. Rev. Water* **2020**, *7*, e1407. [CrossRef]
8. Dietz, K.J.; Zörb, C.; Geilfus, C.M. Drought and crop yield. *Plant Biol.* **2021**, *23*, 881–893. [CrossRef]
9. Moreno-Jiménez, E.; Plaza, C.; Saiz, H.; Manzano, R.; Flagmeier, M.; Maestre, F.T. Aridity and reduced soil micronutrient availability in global drylands. *Nat. Sustain.* **2019**, *2*, 371–377. [CrossRef]
10. Borges, C.E.; Von dos Santos Veloso, R.; da Conceição, C.A.; Mendes, D.S.; Ramirez-Cabral, N.Y.; Shabani, F.; Shafapourtehrany, M.; Nery, M.C.; da Silva, R.S. Forecasting *Brassica napus* production under climate change with a mechanistic species distribution model. *Sci. Rep.* **2023**, *13*, 12656. [CrossRef]
11. Zdziennicka, A.; Szymczyk, K.; Jańczuk, B.; Longwic, R.; Sander, P. Surface, volumetric, and wetting properties of oleic, linoleic, and linolenic acids with regards to application of canola oil in diesel engines. *Appl. Sci.* **2019**, *9*, 3445. [CrossRef]
12. Zhu, M.; Monroe, J.G.; Suhail, Y.; Villiers, F.; Mullen, J.; Pater, D.; Hauser, F.; Jeon, B.W.; Bader, J.S.; Kwak, J.M.; et al. Molecular and systems approaches towards drought-tolerant canola crops. *New Phytol.* **2016**, *210*, 1169–1189. [CrossRef] [PubMed]
13. Chunjie, Y.; Xuekun, Z.; Chongshun, Z. Effects of drought simulated by PEG-6000 on germination and seedling growth of rapeseed (*Brassica napus* L.). *Chin. J. Oil Crop Sci.* **2007**, *29*, 425–430.
14. Michalak, M.; Antoszewski, M.; Kamiński, D.; Mierek-Adamski, A.; Dąbrowska, B.G. Priming of *Brassica napus* L. seeds with aqueous extract from mistletoe (*Viscum album* L.) boosts the content of photosynthetic pigments. *Ecol. Quest.* **2024**, *35*, 1–15.
15. Paparella, S.; Araújo, S.S.; Rossi, G.; Wijayasinghe, M.; Carbonera, D.; Balestrazzi, A. Seed priming: State of the art and new perspectives. *Plant Cell Rep.* **2015**, *34*, 1281–1293. [CrossRef]
16. El-Saadony, M.T.; Saad, A.M.; Soliman, S.M.; Salem, H.M.; Ahmed, A.I.; Mahmood, M.; El-Tahan, A.M.; Ebrahim, A.A.M.; Abd El-Mageed, T.A.; Negm, S.H.; et al. Plant growth-promoting microorganisms as biocontrol agents of plant diseases: Mechanisms, challenges and future perspectives. *Front. Plant Sci.* **2022**, *13*, 923880. [CrossRef]
17. Antoszewski, M.; Mierek-Adamska, A.; Dąbrowska, G.B. The importance of microorganisms for sustainable agriculture—A review. *Metabolites* **2022**, *12*, 1100. [CrossRef]
18. Dąbrowska, G.B.; Zdziechowska, E. The role of rhizobacteria in the stimulation of the growth and development processes and protection of plants against environmental factors. *Prog. Plant Prot.* **2015**, *55*, 498–506.
19. Saravanakumar, K.; Fan, L.; Fu, K.; Yu, C.; Wang, M.; Xia, H.; Sun, J.; Li, Y.; Chen, J. Cellulase from *Trichoderma harzianum* interacts with roots and triggers induced systemic resistance to foliar disease in maize. *Sci. Rep.* **2016**, *6*, 35543. [CrossRef]
20. Patel, P.; Gajjar, H.; Joshi, B.; Krishnamurthy, R.; Amaresan, N. Inoculation of salt-tolerant *Acinetobacter* sp. (RSC9) improves the sugarcane (*Saccharum* sp. Hybrids) growth under salinity stress condition. *Sugar Tech* **2022**, *24*, 494–501. [CrossRef]

21. Radhakrishnan, R.; Khan, A.L.; Kang, S.M.; Lee, I.J. A comparative study of phosphate solubilization and the host plant growth promotion ability of *Fusarium verticillioides* RK01 and *Humicola* sp. KNU01 under salt stress. *Ann. Microbiol.* **2015**, *65*, 585–593. [CrossRef]
22. Zahir, Z.A.; Ghani, U.; Naveed, M.; Nadeem, S.M.; Asghar, H.N. Comparative effectiveness of *Pseudomonas* and *Serratia* sp. containing ACC-deaminase for improving growth and yield of wheat (*Triticum aestivum* L.) under salt-stressed conditions. *Arch. Microbiol.* **2009**, *191*, 415–424. [CrossRef] [PubMed]
23. Dąbrowska, G.B.; Garstecka, Z.; Trejgell, A.; Dąbrowski, H.P.; Konieczna, W.; Szyp-Borowska, I. The impact of forest fungi on promoting growth and development of *Brassica napus* L. *Agronomy* **2021**, *11*, 2475. [CrossRef]
24. Sharma, M.; Mishra, V.; Rau, N.; Sharma, R.S. Increased iron-stress resilience of maize through inoculation of siderophore-producing *Arthrobacter globiformis* from mine. *J. Basic Microbiol.* **2016**, *56*, 719–735. [CrossRef] [PubMed]
25. Garstecka, Z.; Antoszewski, M.; Mierek-Adamska, A.; Krauklis, D.; Niedojadło, K.; Kaliska, B.; Hrynkiewicz, K.; Dąbrowska, G.B. *Trichoderma viride* colonizes the roots of *Brassica napus* L., alters the expression of stress-responsive genes, and increases the yield of canola under field conditions during drought. *Int. J. Mol. Sci.* **2023**, *24*, 15349. [CrossRef] [PubMed]
26. Flores, A.C.; Luna, A.A.E.; Portugal, V.O. Yield and quality enhancement of marigold flowers by inoculation with *Bacillus subtilis* and *Glomus fasciculatum*. *J. Sustain. Agric.* **2007**, *31*, 21–31. [CrossRef]
27. Janowska, B.; Andrzejak, R.; Kosiada, T. The influence of fungi of the *Trichoderma* genus on the flowering of *Freesia refracta* Klatt "Argentea" in winter. *Hortic. Sci.* **2020**, *47*, 203–210. [CrossRef]
28. Khan, A.L.; Hamayun, M.; Kim, Y.H.; Kang, S.M.; Lee, J.H.; Lee, I.J. Gibberellins producing endophytic *Aspergillus fumigatus* sp. LH02 influenced endogenous phytohormonal levels, isoflavonoids production and plant growth in salinity stress. *Process Biochem.* **2011**, *46*, 440–447. [CrossRef]
29. Parmasi, Z.; Tahmasebi, Z.; Zare, M.J.; Nourollahi, K.; Kanouni, H. Biocontrol of Ascochyta blight by *Azospirillum* sp. depending on the degree of resistance of chickpea genotypes. *J. Phytopathol.* **2019**, *167*, 601–607. [CrossRef]
30. Turkan, S.; Mierek-Adamska, A.; Kulasek, M.; Konieczna, W.B.; Dabrowska, G.B. New seed coating containing *Trichoderma viride* with anti-pathogenic properties. *PeerJ* **2023**, *11*, e15392. [CrossRef]
31. Rojas-Tapias, D.F.; Bonilla, R.; Dussán, J. Effect of inoculation and co-inoculation of *Acinetobacter* sp. RG30 and *Pseudomonas putida* GN04 on growth, fitness, and copper accumulation of maize (*Zea mays*). *Water Air Soil Pollut.* **2014**, *225*, 2232. [CrossRef]
32. Dąbrowska, G.; Hrynkiewicz, K.; Trejgell, A.; Baum, C. The effect of plant growth-promoting rhizobacteria on the phytoextraction of Cd and Zn by *Brassica napus* L. *Int. J. Phytoremediat.* **2017**, *19*, 597–604. [CrossRef] [PubMed]
33. Szymańska, S.; Dąbrowska, G.B.; Tyburski, J.; Niedojadło, K.; Piernik, A.; Hrynkiewicz, K. Boosting the *Brassica napus* L. tolerance to salinity by the halotolerant strain *Pseudomonas stutzeri* ISE12. *Environ. Exp. Bot.* **2019**, *163*, 55–68. [CrossRef]
34. Bano, Q.; Ilyas, N.; Bano, A.; Zafar, N.; Akram, A.; Fayaz, A.; Hassan, U.L. Effect of *Azospirillum* inoculation on maize (*Zea mays* L.) under drought stress. *Pak. J. Bot.* **2013**, *45*, 13–20.
35. Rubin, R.L.; van Groenigen, K.J.; Hungate, B.A. Plant growth promoting rhizobacteria are more effective under drought: A meta-analysis. *Plant Soil* **2017**, *416*, 309–323. [CrossRef]
36. Zhao, X.; Yuan, X.; Xing, Y.; Dao, J.; Zhao, D.; Li, Y.; Li, W.; Wang, Z. A meta-analysis on morphological, physiological and biochemical responses of plants with PGPR inoculation under drought stress. *Plant Cell Environ.* **2023**, *46*, 199–214. [CrossRef]
37. Skrzypczak, D.; Mikula, K.; Kossińska, N.; Widera, B.; Warchoł, J.; Moustakas, K.; Chojnacka, K.; Witek-Krowiak, A. Biodegradable hydrogel materials for water storage in agriculture-review of recent research. *Desalination Water Treat.* **2020**, *194*, 324–332. [CrossRef]
38. Liu, Y.; Wang, J.; Chen, H.; Cheng, D. Environmentally friendly hydrogel: A review of classification, preparation and application in agriculture. *Sci. Total Environ.* **2022**, *846*, 157303. [CrossRef]
39. Zhu, J.; Suhaimi, F.; Lim, J.Y.; Gao, Z.; Swarup, S.; Loh, C.S.; Li, J.; Ong, C.N.; Tan, W.K. A field study on using soybean waste-derived superabsorbent hydrogel to enhance growth of vegetables. *Sci. Total Environ.* **2022**, *851*, 158141. [CrossRef]
40. Makowska, M.; Borowski, E.; Ziemba, A. The gas exchange and yielding of strawberry plants cultivated in black soil and sandy soil with the addition of hydrogel. *Acta Sci. Pol. Hortorum Cultus* **2005**, *4*, 153–161.
41. Mierek-Adamska, A.; Kotowicz, K.; Goc, A.; Boniecka, J.; Berdychowska, J.; Dąbrowska, G.B. Potential involvement of rapeseed (*Brassica napus* L.) metallothioneins in the hydrogen peroxide-induced regulation of seed vigour. *J. Agron. Crop Sci.* **2019**, *205*, 598–607. [CrossRef]
42. Daudi, A.; O'Brien, J. Detection of hydrogen peroxide by DAB staining in *Arabidopsis* leaves. *Bio-Protocol* **2012**, *2*, e263. [CrossRef] [PubMed]
43. R Core Team. *R: A Language and Environment for Statistical Computing*; PBC: Boston, MA, USA, 2024.
44. Pandey, A.; Yarzábal, L.A. Bioprospecting cold-adapted plant growth promoting microorganisms from mountain environments. *Appl. Microbiol. Biotechnol.* **2018**, *103*, 643–657. [CrossRef] [PubMed]
45. Van Elsas, J.D.; Chiurazzi, M.; Mallon, C.A.; Elhottova, D.; Krištůfek, V.; Salles, J.F. Microbial diversity determines the invasion of soil by a bacterial pathogen. *Proc. Natl. Acad. Sci. USA* **2012**, *109*, 1159–1164. [CrossRef] [PubMed]

46. Mathes, F.; Murugaraj, P.; Bougoure, J.; Pham, V.T.H.; Truong, V.K.; Seufert, M.; Wissemeier, A.H.; Mainwaring, D.E.; Murphy, D.V. Engineering rhizobacterial community resilience with mannose nanofibril hydrogels towards maintaining grain production under drying climate stress. *Soil Biol. Biochem.* **2020**, *142*, 107715. [CrossRef]
47. Janczak, K.; Hrynkiewicz, K.; Znajewska, Z.; Dąbrowska, G. Use of rhizosphere microorganisms in the biodegradation of PLA and PET polymers in compost soil. *Int. Biodeterior. Biodegrad.* **2018**, *130*, 65–75. [CrossRef]
48. Janczak, K.; Dąbrowska, G.B.; Raszkowska-Kaczor, A.; Kaczor, D.; Hrynkiewicz, K.; Richert, A. Biodegradation of the plastics PLA and PET in cultivated soil with the participation of microorganisms and plants. *Int. Biodeterior. Biodegrad.* **2020**, *155*, 105087. [CrossRef]
49. Dąbrowska, G.B.; Tylman-Mojżeszek, W.; Mierek-Adamska, A.; Richert, A.; Hrynkiewicz, K. Potential of *Serratia plymuthica* IV-11-34 strain for biodegradation of polylactide and poly(ethylene terephthalate). *Int. J. Biol. Macromol.* **2021**, *193*, 145–153. [CrossRef]
50. Proença, D.N.; Schwab, S.; Vidal, M.S.; Baldani, J.I.; Xavier, G.R.; Morais, P.V. The nematicide *Serratia plymuthica* M24T3 colonizes *Arabidopsis thaliana*, stimulates plant growth, and presents plant beneficial potential. *Braz. J. Microbiol.* **2019**, *50*, 777–789. [CrossRef]
51. Abuamsha, R.; Salman, M.; Ehlers, R.-U. Effect of seed priming with *Serratia plymuthica* and *Pseudomonas chlororaphis* to control *Leptosphaeria maculans* in different oilseed rape cultivars. *Eur. J. Plant Pathol.* **2011**, *130*, 287–295. [CrossRef]
52. Adam, E.; Müller, H.; Erlacher, A.; Berg, G. Complete genome sequences of the *Serratia plymuthica* strains 3Rp8 and 3Re4-18, two rhizosphere bacteria with antagonistic activity towards fungal phytopathogens and plant growth promoting abilities. *Stand. Genom. Sci.* **2016**, *11*, 61. [CrossRef]
53. Singh, R.P.; Jha, P.N. The multifarious PGPR *Serratia marcescens* CDP-13 augments induced systemic resistance and enhanced salinity tolerance of wheat (*Triticum aestivum* L.). *PLoS ONE* **2016**, *11*, e0155026. [CrossRef] [PubMed]
54. Koo, S.Y.; Cho, K.S. Isolation and characterization of a plant growth-promoting rhizobacterium, *Serratia* sp. SY5. *J. Microbiol. Biotechnol.* **2009**, *19*, 1431–1438. [PubMed]
55. El-Esawi, M.A.; Alaraidh, I.A.; Alsahli, A.A.; Alzahrani, S.M.; Ali, H.M.; Alayafi, A.A.; Ahmad, M. *Serratia liquefaciens* KM4 improves salt stress tolerance in maize by regulating redox potential, ion homeostasis, leaf gas exchange and stress-related gene expression. *Int. J. Mol. Sci.* **2018**, *19*, 3310. [CrossRef]
56. Patel, S.K.; Singh, S.; Benjamin, J.C.; Singh, V.R.; Bisht, D.; Lal, R.K. Plant growth-promoting activities of *Serratia marcescens* and *Pseudomonas fluorescens* on *Capsicum annuum* L. plants. *Ecol. Front.* **2024**, *44*, 654–663. [CrossRef]
57. Dastager, S.G.; Deepa, C.K.; Pandey, A. Potential plant growth-promoting activity of *Serratia nematodiphila* NII-0928 on black pepper (*Piper nigrum* L.). *World J. Microbiol. Biotechnol.* **2011**, *27*, 259–265. [CrossRef]
58. Kotoky, R.; Nath, S.; Kumar Maheshwari, D.; Pandey, P. Cadmium resistant plant growth promoting rhizobacteria *Serratia marcescens* S2I7 associated with the growth promotion of rice plant. *Environ. Sustain.* **2019**, *2*, 135–144. [CrossRef]
59. Niu, H.; Sun, Y.; Zhang, Z.; Zhao, D.; Wang, N.; Wang, L.; Guo, H. The endophytic bacterial entomopathogen *Serratia marcescens* promotes plant growth and improves resistance against *Nilaparvata* lugens in rice. *Microbiol. Res.* **2022**, *256*, 126956. [CrossRef]
60. Mahdi, I.; Hafidi, M.; Allaoui, A.; Biskri, L. Halotolerant endophytic bacterium *Serratia rubidaea* ED1 enhances phosphate solubilization and promotes seed germination. *Agriculture* **2021**, *11*, 224. [CrossRef]
61. Wang, S.; Ouyang, L.; Ju, X.; Zhang, L.; Zhang, Q.; Li, Y. Survey of plant drought-resistance promoting bacteria from *Populus euphratica* tree living in arid area. *Indian J. Microbiol.* **2014**, *54*, 419–426. [CrossRef]
62. Pan, B.; Vessey, J.K.; Smith, D.L. Response of field-grown soybean to co-inoculation with the plant growth promoting rhizobacteria *Serratia proteamaculans* or *Serratia liquefaciens*, and *Bradyrhizobium japonicum* pre-incubated with genistein. *Eur. J. Agron.* **2002**, *17*, 143–153. [CrossRef]
63. Han, H.; Sheng, X.; Hu, J.; He, L.; Wang, Q. Metal-immobilizing *Serratia liquefaciens* CL-1 and *Bacillus thuringiensis* X30 increase biomass and reduce heavy metal accumulation of radish under field conditions. *Ecotoxicol. Environ. Saf.* **2018**, *161*, 526–533. [CrossRef] [PubMed]
64. Tariq, Z.; Iqbal, D.N.; Rizwan, M.; Ahmad, M.; Faheem, M.; Ahmed, M. Significance of biopolymer-based hydrogels and their applications in agriculture: A review in perspective of synthesis and their degree of swelling for water holding. *RSC Adv.* **2023**, *13*, 24731–24754. [CrossRef] [PubMed]
65. Guilherme, M.R.; Aouada, F.A.; Fajardo, A.R.; Martins, A.F.; Paulino, A.T.; Davi, M.F.T.; Rubira, A.F.; Muniz, E.C. Superabsorbent hydrogels based on polysaccharides for application in agriculture as soil conditioner and nutrient carrier: A review. *Eur. Polym. J.* **2015**, *72*, 365–385. [CrossRef]
66. Sahmat, S.S.; Rafii, M.Y.; Oladosu, Y.; Jusoh, M.; Hakiman, M.; Mohidin, H. A systematic review of the potential of a dynamic hydrogel as a substrate for sustainable agriculture. *Horticulturae* **2022**, *8*, 1026. [CrossRef]
67. Chen, X.; Huang, L.; Mao, X.; Liao, Z.; He, Z. A comparative study of the cellular microscopic characteristics and mechanisms of maize seedling damage from superabsorbent polymers. *Pedosphere* **2017**, *27*, 274–282. [CrossRef]

68. Dhanalakshmi, A.; Vijayakumari, K.K.; Marimuthu, S.; Surendran, U. Evaluation of different soil textures in combination with growing media on growth, yield, and water productivity of blackgram. *Commun. Soil Sci. Plant Anal.* **2020**, *51*, 2670–2682.
69. Al Rohily, K.; El-Hamshary, H.; Ghoneim, A.; Modaihsh, A. Controlled release of phosphorus from superabsorbent phosphate-bound alginate-graft-polyacrylamide: Resistance to soil cations and release mechanism. *ACS Omega* **2021**, *5*, 32919–32929. [CrossRef]
70. Wu, Y.; Li, S.; Chen, G. Hydrogels as water and nutrient reservoirs in agricultural soil: A comprehensive review of classification, performance, and economic advantages. *Environ. Dev. Sustain.* **2023**, *26*, 24653–24685. [CrossRef]
71. Womack, N.C.; Piccoli, I.; Camarotto, C.; Squartini, A.; Guerrini, G.; Gross, S.; Maggini, M.; Cabrera, M.L.; Morari, F. Hydrogel application for improving soil pore network in agroecosystems. Preliminary results on three different soils. *CATENA* **2022**, *208*, 105759. [CrossRef]
72. Shen, S.S.; Kim, W.-I.; Park, C.S. Effect of hydrogel on survival of *Serratia plymuthica* A21-4 in soils and plant disease suppression. *Plant Pathol. J.* **2006**, *22*, 364–368. [CrossRef]
73. Erdogan, S.; Kartal, M.T.; Pata, U.K. Does climate change cause an upsurge in food prices? *Foods* **2024**, *13*, 154. [CrossRef]
74. Schaub, S.; Benni, N.E. How do price (risk) changes influence farmers' preferences to reduce fertilizer application? *Agric. Econ.* **2024**, *55*, 365–383. [CrossRef]

*agriculture*

MDPI

*Article*

# The Presence of Arbuscular Mycorrhizal Fungi in the Rhizosphere of Transgenic Rapeseed Overexpressing a *Trichoderma Thkel1* Gene Improves Plant Development and Yield

Carlos Nicolás [1,*,†], Mónica Calvo-Polanco [1,†], Jorge Poveda [2,*], Ana Alonso-Ramírez [3], Julio Ascaso [4], Vicent Arbona [5] and Rosa Hermosa [4]

1 Department of Botany and Plant Physiology, Institute for Agribiotechnology Research (CIALE), University of Salamanca, 37185 Salamanca, Spain; mcalvopolanco@usal.es
2 Department of Plant Production and Forest Resources, University Institute for Research in Sustainable Forest Management (iuFOR), University of Valladolid, 34004 Palencia, Spain
3 Independent Researcher, 37007 Salamanca, Spain
4 Department of Microbiology and Genetics, Institute for Agribiotechnology Research (CIALE), University of Salamanca, 37185 Salamanca, Spain; julioascaso@usal.es (J.A.); rhp@usal.es (R.H.)
5 Department of Agricultural and Environmental Sciences, Jaume I University, 12071 Castellón, Spain; arbona@uji.es
* Correspondence: cnicolas@usal.es (C.N.); jorge.poveda@uva.es (J.P.); Tel.: +34-923294500 (ext. 5107) (C.N.)
† These authors contributed equally to this work.

**Abstract:** Most of the plants belonging to the family of Brassicaceae are non-hosts for arbuscular mycorrhizal fungi (AMF). These plants are known to produce glucosinolates (GSL), a group of allelopathic compounds, with a role in plant defense. The overexpression of the *Thkel1* from *Trichoderma harzianum* in rapeseed (BnKel) plants, this gene encoding a protein that shares similarities with Brassicaceae plant's nitrile-specifier and epithiospecifier proteins, modified GSL metabolism, reducing the accumulation of toxic isothiocyanates due to hydrolysis of these secondary metabolites. Here, we have analyzed the effect of AMF application on the GSL profiles and the development and yield of BnKel plants. Our results showed that the reduction of GSL compounds on transgenic plants was not enough to allow the formation of arbuscules and vesicles characteristics of an AMF mycorrhizal association. However, the inoculation of transgenic rapeseed plants expressing *Thkel1* with AMF improved seed yield and fatty acid composition of the oilseed, showing a beneficial effect of AMF in these plants. The achievement of this effective beneficial association among mycorrhizas and rapeseed plants opens new opportunities in agribiotechnology for the use of AMF as biofertilizers in Brassicaceae crops with potential application in medical, animal and industrial biotechnology.

**Keywords:** AMF; Brasicaceae; glucosinolates; rapeseed; *Thkel1*

## 1. Introduction

The genus *Brassica* is probably the most agriculturally relevant group of Brassicaceae, from an agronomic point of view, as it includes more than 30 species and hybrids with economic interest and worldwide distribution [1]. Members of the family Brassicaceae are known to produce glucosinolates (GSL), allelopathic compounds related to plant defense [2–4]. These compounds are hydrolyzed by thioglucosidases, also known as myrosinases, to form toxic isothiocyanates, playing a role in plant defense against bacteria, fungi, and insect herbivores, while producing other compounds with different biological activities [3]. The GSL breakdown begins when the spatial separation of GSL and myrosinases is suppressed after tissular damage induced by an attacker. Other groups of proteins, called nitrile-specifier (NSP) and epithiospecifier proteins (ESP) can interact with myrosinases, addressing the reaction to the formation of nitriles or epithionitriles, compounds with less toxicity than the GSL-derived isothiocyanates [3,5].

Beneficial mutualistic symbiosis among plants and fungi are critical features for plant survival and development. In this context, *Trichoderma* and mycorrhizal fungi are of interest for eco-sustainable agriculture [6,7]. *Trichoderma* includes soil-borne filamentous fungi that are used as direct biocontrol agents against phytopathogens and in their communication with the plant can induce its growth and prime defense responses to pathogens and environmental stresses [7–9]. In addition, several studies have shown that *Trichoderma* constitutes an important source of genes to generate genetically modified (GM) crops [10]. On the other side, arbuscular mycorrhizal fungi (AMF) are characterized by the formation of arbuscules within the root cells, in these symbiotic associations, the presence of AMF improves water and nutrient uptake for these plants, while the fungus receives photosynthetic carbon in the form of sugars and fatty acids [11,12]. Considering crop species, AMF symbiosis is an outstanding feature from an economic point of view. However, the loss of key genes to establish symbiosis during evolution in some flowering plants, as those of the Brasicaceace family [13], has led them to the inability to form AM symbiotic associations [14]. Recently, it has been suggested that the presence of AMF is not entirely impaired in the non-fungal host *Arabidopsis* [15], and although it has been reported the development of rudimentary arbuscular mycorrhizal (AM) phenotypes in some Brasicaceae species, it seems clear that it is not a true AMF host [16,17].

In a previous work, we characterized the *Thkel1* gene from *Trichoderma harzianum*, encoding the ThKEL1 protein that contains five repeated Kelch domains and sharing similarity to NSPs and ESPs of plant origin [18]. Kelch domain proteins are key in the regulation of GSL metabolism in plants and several other processes [19]. The overexpression of the *Thkel1* gene in *Arabidopsis* and rapeseed plants led to improved tolerance to abiotic stress conditions and resistance against the leaf pathogens *Botrytis cinerea* and *Phoma lingam*, respectively, while facilitating their root colonization by *Trichoderma* and increasing the production of seeds [18,20]. The aim of the present work was to test whether the reduction of the GSL levels in BnKel plants would allow the formation of arbuscular structures within the roots of these transgenic plants when AMF is applied. Once the establishment of this symbiotic association was verified, the main objective of this work was to test its beneficial effects on seed yielding as well as its potential applications in agrobiotechnology.

## 2. Materials and Methods

### 2.1. Plant Material

*B. napus* cv. Jura and its BnKel1 and BnKel2 transgenic lines expressing the *Thkel1* gene from *T. harzianum* T34, previously described [20], were the plants used in this study. The two independent *Thkel1* transgenic lines exhibited similar phenotypes and most of the results shown throughout the manuscript correspond to BnKel2 transgenic line.

### 2.2. AMF Material

We used, as AM mycorrhizal inoculum, the Miratext-02 formulation (Mirat Fertilizantes, Salamanca, Spain), which contained at least $1 \times 10^6$ spore $kg^{-1}$ of five different AMF species: *Glomus microagregatum*, *Funneliformis mosseae*, *Claroideoglomus claroideum*, *Rhizophagus irregularis* and *R. fasciculatus*. Non-inoculated (control) plants received a 5 mL aliquot consisting of a filtrate from the arbuscular mycorrhizal inoculum (<20 μm) to provide a general microbial population free of arbuscular mycorrhizal propagules.

### 2.3. AMF Inoculation and Plant Growth Conditions

Seeds were germinated in plates on Murashige and Skoog (MS) medium (Duchefa, Haarlem, The Netherlands) complemented with 1% agar and 1% sucrose. Seedlings were grown for seven days under controlled environmental conditions in a growth chamber at 22 °C, 40% relative humidity (RH), and a 16/8 h (day/night) photoperiod at 80–100 $\mu E\ m^{-2}\ s^{-1}$. After that, seedlings were transferred into 5 L pots, containing an autoclaved mixture of peat/vermiculite (3:1), and maintained under greenhouse conditions as previously described [20] and watered as needed. AM inoculum was applied to each pot by burying 1 g of Miratext-02 (1000 spore $g^{-1}$) at 5 cm below the substrate surface just before transplanting the seedlings.

Ten weeks after transplanting (during the formation of floral primordia), roots were separated from the shoots, carefully washed and immediately frozen with liquid nitrogen and pulverized with a mortar. Roots from five plants per condition were pooled and root pools from three biological replicates were considered for statistical analysis. Dry weight was determined in the aerial part of 8-week-old plants. They were kept at 65 °C for 48 h and weighed. Plants of three biological replicates were analyzed for each assayed condition, and five plants were included per replicate.

Siliques were collected at the end of the life cycle (19 weeks) and counted. Three biological replicates were considered, and 15 plants were used per biological replicate and condition.

### 2.4. Visualization of the AMF Structures in Rapeseed Roots

The *B. napus* roots were immersed in a solution of 10% KOH at 90 °C for 10 min, in order to increase the permeability of their cell walls. Subsequently, three washes were carried out with distilled water to eliminate the KOH and one more with a solution of acetic acid at 2% ($v/v$). The roots were then transferred to a solution with the Sheaffer Skrip Ink [Cult Pens, Tiverton, UK (5% ink in 2% acetic acid)] for another 10 min. Finally, the excess ink was discarded, and the roots were rinsed with distilled water, where they were kept until their observation [21]. Pictures were taken with a stereoscopic microscope, Leica M205 FA, equipped with a Leica DFC 495 camera, (Leica, Madrid, Spain).

The (WGA)-Alexa Fluor 488 conjugate (Molecular Probes, Eugene, OR, USA) was used for detailed images of the fungal morphology at the roots. Root sections were cleared with KOH as explained before, treated with HCI 0.1N for 30 min, and infiltrated with 2.5 μg $ml^{-1}$ wheat germ agglutinin in phosphate-buffered saline (PBS 1Xfor 15 min in the dark. Z-stack images were obtained using a laser-scanning confocal fluorescence microscope (Nikon).

### 2.5. DNA- Detection of AMF at the Roots

The presence of AMF DNA in the roots of rapeseed plants was evaluated by real-time quantitative PCR (qPCR) as previously described [20]. The primers 18S-F and 18S-R were used to amplify the 18S rRNA gene of AMF [18S-F: CTTTCGATGGTAGGATAGAGG; 18S-R: ACAACTTTAATATACGCTATTGGA [22], and the primers BnAct-F and BnAct-R were used to amplify the *actin* gene of rapeseed, as an internal reference gene, since it is considered a housekeeping gene [BnAct-F: CCCTGGAATTGCTGACCGTA; BnAct-R: TGGAAAGTGCTGAGGGATGC [23]. Amplifications were performed using an ABI PRISM 7000 Sequence Detection System (Applied Biosystems, Foster City, CA, USA) programmed as previously described [24]. Each PCR was performed in triplicate and DNA from three biological replicates (each one including pooled roots of five plants per condition) was used. Cycle threshold values were used to calculate the amount of fungal DNA using standard curves. Values of AMF DNA were referred to the amount of rapeseed DNA in every corresponding sample.

### 2.6. Phosphorus and Iron Measurement

Phosphorus (P) and iron (Fe) content were determined in the aerial part of 8-week-old plants. All measurements were quantified at the IRNASA analytical service (CSIC, Salamanca, Spain). Five hundred mg of ground samples were used for quantification, and the content of P and Fe was analyzed by means of optical emission spectrometry with source of plasma connected by induction (ICP-OES), using methodology previously described [25]. Plants of three biological replicates were analyzed for each assayed condition, and five plants were included per replicate. Results are expressed in mg $g^{-1}$ dry weight for P and mg $kg^{-1}$ dry weight for Fe.

### 2.7. GintPT and GintAMT2 Gene Expression Analysis by qPCR

Total RNA was isolated from roots using TRIZOL® reagent (Invitrogen Life Technologies, Carlsbad, CA, USA) following the manufacturer's instructions. Thus, RNA samples were treated with DNase I (Fermentas, Burlington, Canada) and purified using GeneJET RNA Cleanup and Concentration Micro Kit (Thermo Scientific, Waltham, MA, USA). Finally, cDNA was obtained from 1 µg of RNA using the PrimeScriptTM RT reagent mix kit with an oligo (dT) (Takara Inc., Tokyo, Japan).

PCR reactions were performed using the gene-specific primers for AMF transporters *GintPT* and *GintAMT2*, as described in Fernandez et al. [15], and the thermocycler as indicated above. Each 10 µL reaction mixture contained 1 µL of cDNA, 5 µL of Kapa Sybr® Fast (Roche, Basel, Switzerland), 3.4 µL of deionized water, and 0.3 µL of each primer pair at a final concentration of 0.1 mM. The PCR program consisted of 40 cycles: denaturation, 95 °C for 3 s; annealing, 58 °C for *GinPT* and 60 °C for *GintAMT2* for 30 s; extension, 72 °C for 1 s. The specificity of the PCR amplification procedure was checked with a heat-dissociation protocol (from 58 to 95 °C) after the final cycle of the PCR. Relative quantification of specific mRNA levels was performed using the comparative method of Livak and Schmittgen [26]. Expression values were normalized using the housekeeping gene 18S previously described. Three different root RNA samples for each treatment were used for analysis ($n = 3$), with each of them repeated three times. Negative controls without cDNA were used in all the PCR reactions.

### 2.8. GSL and Oil Profiles

#### 2.8.1. Sample Extraction

Polar and non-polar fractions of oil seeds. Seeds were weighed (c.a. 50 mg, the actual amount was recorded) and crushed using a ball mill and glass beads. The resulting oily paste was thoroughly mixed with 300 µL of pure methanol (LC/MS degree) supplemented with biochanin A at 1 mg $L^{-1}$ and sonicated for 10 min at room temperature. After centrifugation at 10,000 rpm for 10 min at 4 °C, the supernatant was recovered and combined with 400 µL of ultrapure water and 200 µL of chloroform. The upper water layer was recovered for polar metabolite analysis whereas the lower organic layer was dried down under vacuum and subsequently reconstituted in pure *n*-butanol (LC/MS degree) for non-polar metabolite analysis described below.

GSL and other semipolar compounds. Freeze-dried root samples (c.a. 10 mg, actual amount was recorded), were extracted in 500 µL of 70% aqueous methanol supplemented with biochanin A at 1 mg $L^{-1}$ (internal standard, IS) by ultrasonication (10 min) at room temperature. After extraction, samples were centrifuged at 10,000 rpm for 10 min at 4 °C and supernatants were recovered. Prior to LC/ESI-QqTOF-MS analysis, supernatants were filtered through 0.2 µm PTFE syringe filters (Whatman International Inc., Kent, UK). Analyses of polar fractions and GSLs were carried out as described below.

#### 2.8.2. LC/ESI-QqTOF-MS Analyses

Chromatographic separations were performed on a 100 mm × 2.1 mm i.d., 1.6 µm, Luna Omega 1.6u Polar C18 (Phenomenex, Torrance, CA, USA). For non-polar fractions, $H_2O$:cyanuric acid (ACN) (15:85, *v*/*v*) supplemented with 0.01% HCOOH and 0.5 mM $NH_4Ac$ (A) and *n*-butanol containing 0.01% HCOOH and 0.5 mM $NH_4Ac$ (B) were used as solvents. Analysis of GSLs and polar fractions was achieved using $H_2O$ (A) and ACN (B), both supplemented with 0.1% formic acid. In both cases, flow rate was 0.3 mL $min^{-1}$, column temperature was 40 °C and samples were maintained at 12 °C to slow down degradation. The mass spectrometer was operated in both negative and positive electrospray modes. Argon was used as the collision gas, and nitrogen was used as the nebulizer as well as desolvation gas set at 60 and 800 L $h^{-1}$, respectively. Exact mass measurements were provided by monitoring the reference compound lockmass leucine enkephalin. After acquisition, mass chromatograms were converted to netCDF and subsequently processed with xcms and CAMERA packages [27]. Identification of compounds was achieved by retention time and mass spectra comparison with pure standards, when available, or tentatively annotated based on mass spectral matching with those available in public databases (Massbank, HMDB and Metlin, La Jolla, CA, USA) or the literature.

### 2.9. Statistical Analysis

Statistical analysis was carried out with the Statistix 8.0 software. Data were analyzed using one-way ANOVA and the post hoc Tukey's test at $p < 0.05$ (data in tables), or by two-way ANOVA followed by Sidak's multiple comparison test at * $p < 0.05$; ** $p < 0.01$; *** $p < 0.001$; **** $p < 0.0001$ (for GLS analyses). For analysis of productivity, P and Fe content, and colonization rates, a one-way ANOVA using Tukey's multiple range test at $p < 0.05$ was used for pairwise comparisons, indicating the significant differences with different letters.

## 3. Results

### 3.1. Presence of AMF in Thkel1-Overexpressing Rapeseed Plants

Wild-type and *Thkel1*-overexpressing (BnKel) rapeseed plants were inoculated with a pre-commercial AMF inoculum (provided by Mirat Fertilizantes, Salamanca, Spain) containing five different AMF species. In transgenic BnKel plants, some rudimentary structures could be observed, although they could not be identified as true arbuscules or vesicles characteristic of AMF (Figure 1). The presence of AMF at the root samples was further detected by qPCR in the two transgenic lines analyzed (Table 1). No fungal rudimentary structure or DNA was detected in wild-type plants (Figure 1 and Table 1).

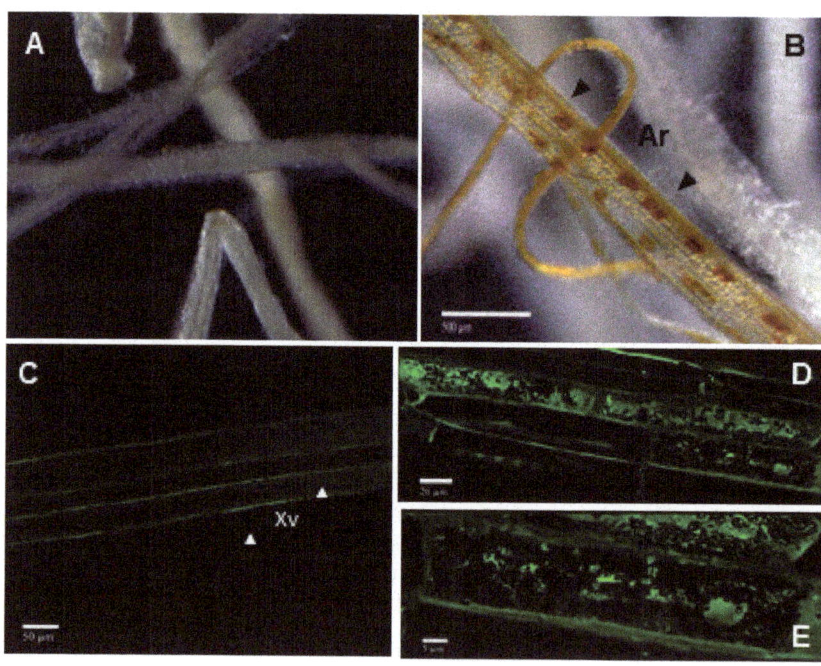


**Figure 1.** Microscopic analyses of the rudimentary fungal structures found in inoculated BnKel2 rapeseed roots. Wild-type (**A**) and BnKel2 roots (**B**) inoculated with AMF were visualized by light microscopy in plants stained with ink and by confocal microscopy in plants stained with the WGA-Alexa Fluor 488 ((**C**), wild-type, and (**D**,**E**), BnKel2 plants). (Ar: arbuscule; Xv, xylem vessels).

**Table 1.** AMF quantification in rapeseed roots of wild-type (WT) and *Thkel1* transgenic lines 1 (BnKel1) and 2 (BnKel2) plants inoculated with a pre-commercial AMF mixture.

| AMF Inoculated Plants | Plant | | | AMF | | | Ratio [3] |
|---|---|---|---|---|---|---|---|
| | Ct | SD | Qty [1] | Ct | SD | Qty [2] | |
| WT | 21.28 | 0.05 | 0.84 | - | 0.0 | 0.0 | 0 [c] |
| BnKel1 | 20.75 | 0.02 | 1.19 | 27.53 | 0.12 | 0.78 | 0.65 ± 0.03 [b] |
| BnKel2 | 20.98 | 0.03 | 0.90 | 26.38 | 0.22 | 0.67 | 0.74 ± 0.04 [a] |

[1] Quantity of plant DNA (ng) referred to rapeseed *actin* gene. [2] Quantity of fungi DNA (ng) referred to AMF 18S rRNA. [3] Proportion of fungal DNA vs. plant DNA. Values represent the mean (±SD) of three biological replicates for each condition ($n = 3$) and five technical replicates each. Statistically significant differences were determined after one-way analysis of variance (ANOVA) followed by Tukey's test and showed with different letters ($p < 0.05$). (-): Absence of amplification.

### 3.2. Gene Expression Analysis of Phosphate and Ammonium Transporters of the AMF *Rhizophagus Irregularis*

We analyzed, in rapeseed roots, the transcript levels of two AMF transporter genes, *GintPT* (encoding a high-affinity phosphate transporter) and *GintAMT2* (encoding a high-affinity ammonium transporter) [15]. The expression of both genes was detected in AMF-BnKel roots, but not in non-inoculated BnKel or wild-type plants inoculated or not with AMF (Figure 2). However, no expression of the two sugar transporter genes, typical of AMF symbiosis, was detected by qPCR (data not shown).

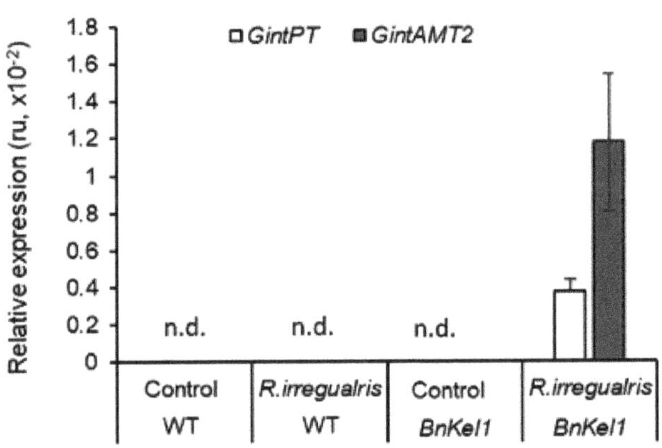


**Figure 2.** AMF *GintPT* and *GintAMT2* transcript levels in inoculated and non-inoculated roots of *Brassica napus* wild-type and BnKel2 transgenic lines. Data are the mean (±SD) of three biological replicates for each type of plant (wild-type and BnKel transgenic lines) and inoculation condition ($n = 3$). Data, obtained from 16-week-old plants, represent the relative quantity (RQ, $2^{-\Delta\Delta Ct}$) of target genes compared to the quantity of the reference gene 18S rRNA. n.d.: not detected.

### 3.3. Changes in P and Fe Content in AMF-BnKel Plants

Compared to the non-inoculated wild-type plants, the levels of the inorganic elements P and Fe decreased in plants when inoculated with AMF. In contrast, the levels of P were higher in the non-inoculated BnKel plants compared to wild-type plants. Both P and Fe contents increased significantly in BnKel plants inoculated with AMF ($p < 0.05$) (Table 2).

### 3.4. Reduction of GSL Content in BnKel Roots Inoculated with AMF

Wild-type plants, inoculated or not with AMF, did not show any significant difference in the levels of indole GSL [indole-3 methyl, (I3M), 4-hydroxy indole-3 methyl (4-HO-I3M), 1-methoxy indole-3 methyl (1-MO-I3M) and 4-methoxy indole-3 methyl (4-MO-I3M) GSL], whereas it was observed a significant increase in the content of aliphatic GSL [4-methyl thio butanal (4-MTB) and 5-methyl thio pentanal (5-MTP) GSL] of AMF-inoculated wild-type plants. The levels of both indole and aliphatic GSL significantly decreased in the roots of BnKel2 plants inoculated with AMF (Figure 3A–F). In addition, a significant increase in the levels of a toxic isothiocyanate [4-methoxy indole-3 methyl ITC (4-MO-I3M-ITC)], derived from the hydrolysis of indole GSL, was detected in AMF-inoculated wild-type plants. However, the lowest content of 4-MO-I3M-ITC was observed in AMF-inoculated BnKel2 plants (Figure 3G).

**Table 2.** Phosphorus (P) and iron (Fe) content in rapeseed leaves, non-inoculated or inoculated with AMF fungi, in wild-type plants (WT or WT + AMF) or plants overexpressing the *Thelk1* gene from *Trichoderma* (BnKel2 or BnKel2 + AMF).

| Plants | P [1] | Fe [2] |
|---|---|---|
| WT | $2.990 \pm 0.063^{c}$ | $82.075 \pm 5.655^{bc}$ |
| WT + AMF | $2.592 \pm 0.242^{d}$ | $71.715 \pm 4.815^{c}$ |
| BnKel2 | $3.715 \pm 0.017^{b}$ | $86.218 \pm 7.765^{b}$ |
| BnKel2 + AMF | $4.994 \pm 0.496^{a}$ | $138.379 \pm 18.993^{a}$ |

Wild-type (WT) and *Thkel1* transgenic line 2 (BnKel2) inoculated with AMF (+AMF). Values represent the mean (±SD) of three biological replicates for each condition ($n = 3$) and three technical replicates each. Different letters in the same column indicate significative differences among treatment means according to Tukey's test ($p < 0.05$). [1] Data in mg $g^{-1}$ of plant tissue. [2] Data in mg $kg^{-1}$ of plant tissue.

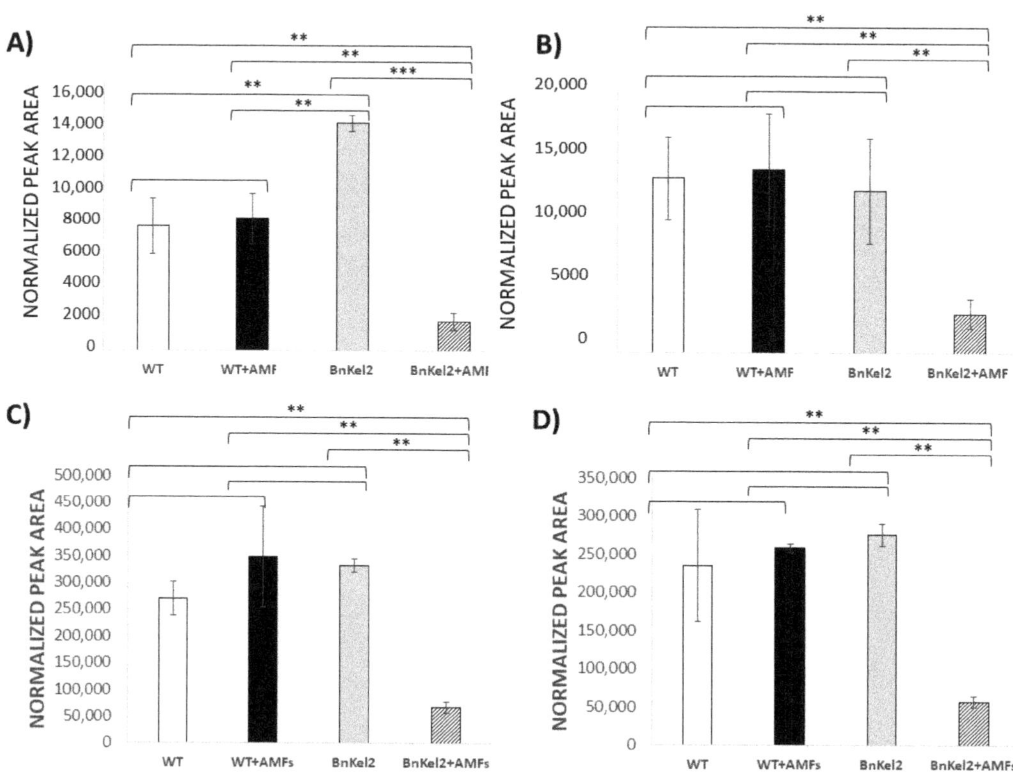


**Figure 3.** *Cont.*

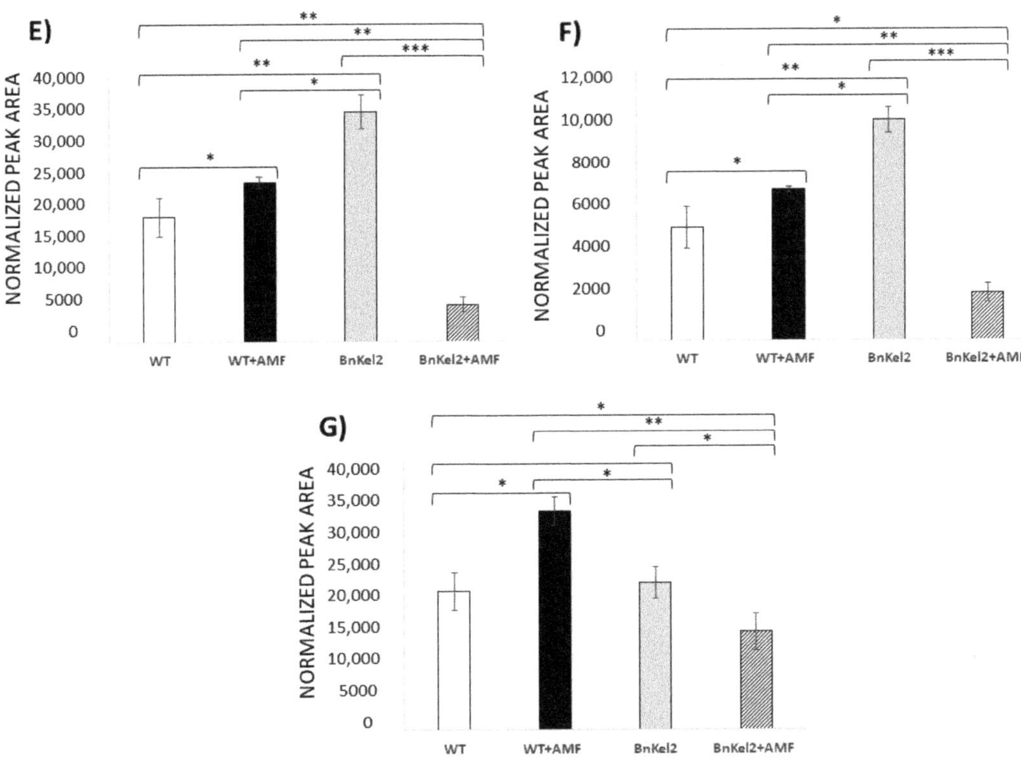


**Figure 3.** Relative quantification of the GSL present in rapeseed roots. (**A**) Indole-3 methyl (I3M); (**B**) 4-hydroxy indole-3 methyl (4-HO-I3M); (**C**) 1-methoxy indole-3 methyl (1-MO-I3M); (**D**) 4-methoxy indole-3 methyl (4-MO-I3M); (**E**) 4-methyl thio butanal (4-MTB); (**F**) 5-methyl thio pentanal (5-MTP); (**G**) 4-4-methoxy indole-3 methyl ITC (4-MO-I3M-ITC). Wild-type (WT) and *Thkel1* transgenic line 2 (BnKel2) inoculated with AMF (+AMF). Values represent the means (±SD) of three biological replicates for each condition ($n = 3$) and five technical replicates each. Two-way analysis of variance (ANOVA) was performed, followed by Sidak's multiple comparison test, indicating significant differences among treatment means as follows: * $p < 0.05$; ** $p < 0.01$; *** $p < 0.001$.

### *3.5. Reduction of GSL Content in BnKel Seeds and Seed Oil Content*

A metabolomic analysis of GSL performed in seeds of wild-type and BnKel2 plants showed a significant increase in the levels of the 4-MO-I3M in seeds of wild-type plants inoculated with AMF. No significant differences were detected in the other GSL determined in wild-type plants. By contrast, a significant decrease in the content of two out of three analyzed GSL was detected in seeds of inoculated BnKel2 plants compared to non-inoculated plants (Figure 4).

Moreover, a significant increase in the levels of octadecatrienoic acid (α-linolenic acid) and glycerophosphocholine was observed in the seeds of BnKel2 plants compared with those of the wild type, regardless of AMF inoculation (Figure 5).

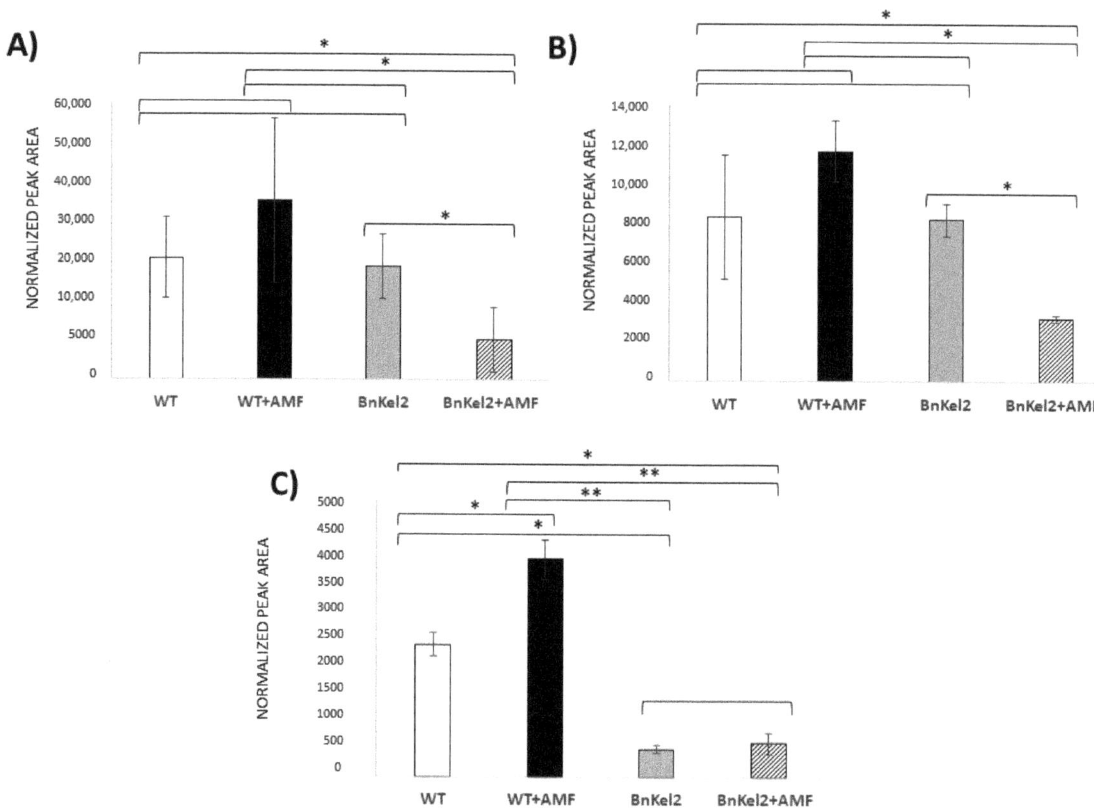


**Figure 4.** Relative quantification of the GSL present in rapeseed seeds. (**A**) 4-methyl thio butanal (4-MTB); (**B**) 5-methyl thio pentanal (5-MTP); (**C**) 4-methoxy indole-3 methyl ITC (4-MO-I3M-ITC). Wild-type (WT) and *Thkel1* transgenic line 2 (BnKel2) inoculated with AMF (+AMF). Values represent the means (±SD) of three biological replicates for each condition ($n = 3$) and five technical replicates each. Two-way analysis of variance (ANOVA) was performed, followed by Sidak's multiple comparison test, indicating significant differences among treatment means as follows: * $p < 0.05$; ** $p < 0.01$.

### 3.6. *Rapeseed Seed Yield*

A significant increase in seed weight was observed in AMF-inoculated BnKel2 plants compared to those non-inoculated. Regardless of AMF inoculation, seed yield was significantly higher in BnKel2 plants compared with that of the wild type (Table 3). Total seed weight per plant increased by about 1.5 times in the AMF-inoculated BnKel2 plants under optimal growth conditions. Further, shoot dry weight increased when BnKel2 plants were inoculated with AMF compared to wild-type or non-inoculated BnKel2 plants (Supplementary Figure S1). These results indicate that the application of AMF to BnKel plants exerted a beneficial effect on rapeseed yield.

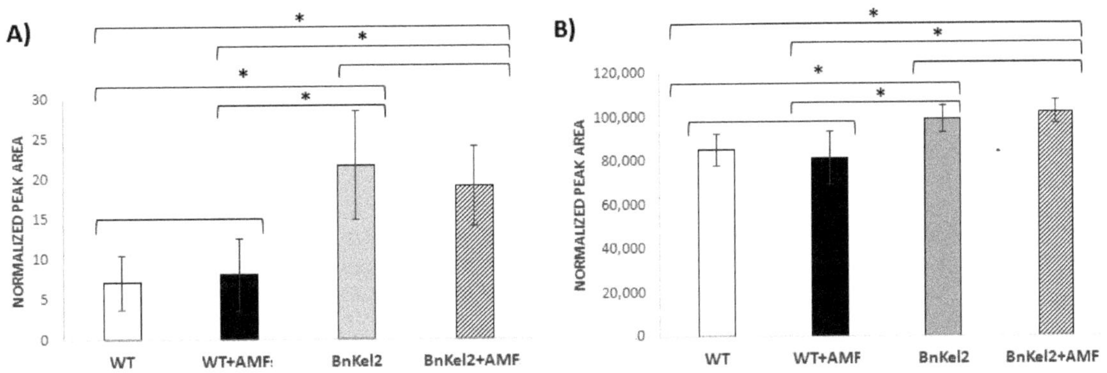


**Figure 5.** Relative quantification of the lipids and lipid-related compounds in the oils of rapeseed seeds. (**A**) Octadecatrienoic acid; (**B**) glycerophosphocholine. Wild-type (WT) and *Thkel1* transgenic line 2 (BnKel2) inoculated with AMF (+AMF). Values represent the means (±SD) of three biological replicates for each condition ($n = 3$), and five technical replicates each. Two-way analysis of variance (ANOVA) was performed, followed by Sidak's multiple comparison test. Asterisks denote significant differences among treatment means at $p \leq 0.05$.

**Table 3.** Seed weight (g) per rapeseed plant under optimal growth conditions (control). Wild-type plants (WT) and plants overexpressing the *Thelk1* gene from *Trichoderma* (BnKel2) were non-inoculated or inoculated with AMF (WT + AMF or BnKel2 + AMF).

| Plants | Seed Weight (g) |
|---|---|
| WT | 61.835 ± 6.226 [c] |
| WT + AMF | 53.653 ± 7.665 [c] |
| BnKel2 | 99.526 ± 9.678 [b] |
| BnKel2 + AMF | 155.596 ± 12.966 [a] |

Values represent the mean (±SD) of three biological replicates for each condition ($n = 3$), and 15 technical replicates each. One-way analysis of variance (ANOVA) was performed, followed by Tukey's test. Different letters indicate significant differences between treatment means at $p < 0.05$.

## 4. Discussion

Many crops, excluding Brassicaceae, can develop mycorrhizal associations that greatly benefit from AMF inoculation [28]. Throughout the evolutionary process, Brassicaceae may have lost their genes related to AMF symbiosis, which seems logical since they were not necessary for their proper development [13]. However, with the present situation of climate change, the need for new technological tools for plant adaptation to adverse environments seems more important than ever. The achievement of an effective beneficial association among mycorrhizas and Brassicaceae plants may be critical in agribiotechnology and opens new opportunities for the use of AMF as biofertilizers in rapeseed crops [6].

Proteins with Kelch domains, including NSP and NSP-like, as ThKEL1, have been associated with the metabolism of GSL and plant defense [2–4,19], and have been also related to the mycorrhization process [24,29,30], in a similar way to that observed in legumes nodulation [31]. The *Thkel1* gene of *T. harzianum* T34, which encodes for an NSP-like protein, increases the fungal tolerance to abiotic stress when it is overexpressed in that strain [18]. Furthermore, when *Thkel1* was overexpressed in Brassicaceae plants, it played a key role in the *T. harzianum* root colonization process [24]. These characteristics of the *Thkel1* gene have made it a candidate of choice to be transferred to *Brassica* plants as a biotechnological tool to modify GSL metabolism and perhaps facilitate the interaction among AMF and these plants. To check this out, we challenged *Thkel1* overexpressing rapeseed plants with a pre-commercial AMF mixture. No fungal DNA was detected in

WT plants (Table 1), but AMF DNA was detected in both BnKel lines, it being indicative that transgenic plants are more sensitive to the presence of AMF symbiotic associations (Figure 1). To support these data, we have obtained similar results regarding the levels of AMF DNA in several independent *Arabidopsis thaliana* transformants, as well as in transgenic rapeseed plants transformed with *Agrobacterium rhizogenes* (Supplementary Table S1) [20]. Some studies have shown that different AMFs are able to form colonization points at the roots of some species of Brassicaceae, but without tissue penetration and nutrient exchange [32,33]. More recently, it has been reported that there is some form of interaction between AMF and non-host *Arabidopsis* plants during the pre-symbiotic stage, whereas the incompatibility appears at later stages, accompanied by an activation of defense responses leading to colonization of the root cortex, but without forming a functional AMF symbiosis [15]. Furthermore, some AMF species can germinate in the presence of roots of wild mustard but without the ability to colonize this Brassicaceae plant [34], which has been found to be due to the formation of allelopathic compounds [35].

Pioneering studies suggested that Brassicaceae plants with low levels of GSL production may establish symbiotic associations with AMF [36]. However, these authors only observed the formation of fungal vesicles in dead cells of the root cortex, concluding that the GSL low content was not the cause of the absence of mycorrhization. Similar results were reported in other ornamental cruciferous plants [37]. Our results show that the application of AMF to BnKel plants increased the P and Fe contents. The higher levels of P content are a known trait in mycorrhizal plants [6,11]. It has been reported that AMF symbiosis can both increase and decrease the Fe uptake, this fact being affected by the fungal species, host plant species and growth conditions [38], where the AMF caused changes in the expression of Fe acquisition and ferric reductase genes [39]. Furthermore, arbuscular formation has been observed in some Brassicaceae species, lacking myrosinase activity and unable to conduct GSL hydrolysis [40]. Another study has also shown that the GSL profile differed among mycorrhizal and non-mycorrhizal Brassicaceae plants. The non-mycorrhizal species accumulated higher levels of the indole GSL 1-MO-I3M and 4-HO-I3M [41], showing a similar profile to that more recently observed in *Brassica juncea* [42]. In addition, that study pointed out that the type of GSL is more important for the process of colonization by AMF than the quantity of GSL generated [41]. Moreover, it has been reported that the indole GSL pathway metabolites provide resistance to AMF mycorrhization in *Arabidopsis* [43,44]. In our study, we have observed a reduction of 1-MO-I3M and 4-HO-I3M levels in AMF-BnKel2 plants, as well as a decrease in the isothiocyanate content derived from one of these compounds (4-MO-I3M-ITC) (Figure 3). Thus, the significant reduction of toxic GSL detected in BnKel2 plants could allow them to establish a symbiotic association with AMF. In any case, this decrease is only detected after root–AMF interaction in BnKel plants, although the *Thkel1* gene could have a role in this interaction since it has been shown to play an important role during *T. harzianum* root colonization in *Arabidopsis* [24].

Recently, the role of GSLs in the control of the crucifer microbiome has been reviewed, highlighting their importance in the absence of effective mycorrhization in this group of plants [44]. Similarly, GSLs are mainly responsible for controlling the fungal diversity of endophytic fungi in crucifers, which is why *Brassica* crops harbor less endophytic fungal diversity in their roots than other taxonomically distant crops [45].

We have also noted a reduction of the GSL content in seeds (Figure 4). The high concentration of GSL in *Brassica* seeds reduces the value of the crop and its usefulness for oil extraction [46]. Different approaches to reduce the GSL content in plants have been attempted, albeit without much success, except that performed by mutation of genes encoding GSL transporters [47]. In addition to increased yield and a reduction of GSL content in seeds from BnKel plants, we have also detected changes in the seed oil composition (Figure 4). In this sense, a significant increase in the levels of octadecatrienoic acid, an unsaturated omega-3 fatty acid with important properties for human nutrition, was detected in seeds harvested from BnKel2 plants. This omega-3 fatty acid is considered to have a positive effect on the prevention of cardiovascular disease and obesity and functions as a neuroprotective agent [48]. Another phospholipid metabolism-related compound whose levels were significantly increased in seeds of BnKel2 was glycerophosphocholine (Figure 5), considered important for treating cognitive impairment, as a supplier of the essential dietary amine choline, a precursor of the neurotransmitter acetylcholine, commonly used as a treatment for Alzheimer´s disease and other dementias [49].

Our results show that although true arbuscules were not developed, the application of AMF to BnKel2 plants significantly increases the following parameters: (i) P and Fe contents (Table 2), (ii) total seed weight per plant (Table 3), and (iii) dry weight of the aerial part of the plant (Supplementary Figure S1). These results agree with the increase in yield and product quality observed in mycorrhizal crops [50], and in Brassicaceae plants after the combined application of AMF and *T. harzianum* [20].

In conclusion, the overexpression of the *Thkel1* gene from the fungus *T. harzianum* in *B. napus* plants allows transgenic plants to benefit from the presence of arbuscular mycorrhizal fungi at their root rhizosphere, mainly by the modification of the GSL content and their hydrolysis products. The final increase in seed yield and the improvement of seed oil composition are among the most remarkable impacts of these *B. napus*–AMF interactions, with an impact on the plant's agronomic value and its potential application in medical, animal and industrial biotechnology.

**Supplementary Materials:** The following supporting information can be downloaded at: https://www.mdpi.com/article/10.3390/agriculture14060851/s1, Figure S1: Growth of 8-weeks old rapeseed plants in the presence or not of AMF; Table S1: AMF quantification in rapeseed and arabidopsis roots of wild type (WT and Col-0, respectively) and *Thkel1* transformed roots with *A. rhizogenes* (BnArKel1) or *A. tumefaciens* (AtKel1) plants inoculated with an AMF mixture.

**Author Contributions:** A.A.-R., J.P., M.C.-P. and J.A. conducted the laboratory work; V.A. performed the metabolomic analysis; C.N., J.P. and M.C.-P. conceived and designed the experiments; C.N., J.P., M.C.-P. and R.H. analyzed the data; C.N., M.C.-P., J.P. and R.H. wrote the paper. All authors have read and agreed to the published version of the manuscript.

**Funding:** This work was supported by the Regional Government of Castile and Leon, Escalera de Excelencia CLU-2018-04 co-funded by the P.O. FEDER 2014-2020, the Projects SA230U13, SA270P18, SA094P20 and IR2020-1-USAL05, and by the Spanish Government, PDI-2021-126575OB-I00 MCIN/AEI/10.13039/501100011033/FEDER, UE.

**Institutional Review Board Statement:** Not applicable.

**Data Availability Statement:** The data generated during this study are available from the corresponding author on reasonable request. Materials will be made available for non-profit research under a material transfer agreement.

**Acknowledgments:** Authors thank Enrique Monte for kindly providing materials and participating in the development of the project. Julio Ascaso thanks the support of his predoctoral contract by Regional Government of Castile and Leon. Mass spectrometric analyses were carried out at Servei Central d'Instrumentació Científica (SCIC) from Universitat Jaume I.

**Conflicts of Interest:** A Spanish patent and an international patent with references P201830755 and PCT/EP2019/081029 have been filled (C.N., J.P., A.A.-R., R.H.). No other author has competing interests.

## References

1. Huang, C.H.; Sun, R.; Hu, Y.; Zeng, L.; Zhang, N.; Cai, L.; Zhang, Q.; Koch, M.A.; Al-Shehbaz, I.; Edger, P.P.; et al. Resolution of *Brassicaceae* phylogeny using nuclear genes uncovers nested radiations and supports convergent morphological evolution. *Mol. Biol. Evol.* **2015**, *33*, 394–412. [CrossRef] [PubMed]
2. Cartea, M.E.; Francisco, M.; Soengas, P.; Velasco, P. Phenolic compounds in *Brassica* vegetables. *Molecules* **2011**, *16*, 251–280. [CrossRef] [PubMed]
3. Angelino, D.; Jeffery, E. Glucosinolate hydrolysis and bioavailability of resulting isothiocyanates: Focus on glucoraphanin. *J. Funct. Foods* **2014**, *7*, 67–76. [CrossRef]
4. Hassan, M.N.; Zainal, Z.; Ismail, I. Plant Kelch containing F-box proteins: Structure, evolution and functions. *RSC Adv.* **2015**, *5*, 42808–42814. [CrossRef]

5. Martínez-Ballesta, M.C.; Carvajal, M. Myrosinase in Brassicaceae: The most important issue for glucosinolate turnover and food quality. *Phytochem. Rev.* **2015**, *14*, 1045–1051. [CrossRef]
6. Berruti, A.; Lumini, E.; Balestrini, R.; Bianciotto, V. Arbuscular mycorrhizal fungi as natural biofertilizers: Let's benefit from past successes. *Front. Microbiol.* **2016**, *6*, 1559. [CrossRef] [PubMed]
7. Woo, S.L.; Hermosa, R.; Lorito, M.; Monte, E. *Trichoderma*: A multipurpose plant beneficial microorganism for eco-sustainable agriculture. *Nat. Rev. Microbiol.* **2023**, *21*, 312–326. [CrossRef] [PubMed]
8. Hermosa, R.; Viterbo, A.; Chet, I.; Monte, E. Plant-beneficial effects of *Trichoderma* and of its genes. *Microbiology* **2012**, *158*, 17–25. [CrossRef] [PubMed]
9. Mendoza-Mendoza, A.; Zaid, R.; Lawry, R.; Hermosa, R.; Monte, E.; Horwitz, B.A.; Mukherjee, P.K. Molecular dialogues between *Trichoderma* and roots: Role of the fungal secretome. *Fungal Biol. Rev.* **2018**, *32*, 62–85. [CrossRef]
10. Nicolás, C.; Hermosa, R.; Rubio, B.; Mukherjee, P.K.; Monte, E. *Trichoderma* genes in plants for stress tolerance-status and prospects. *Plant Sci.* **2014**, *228*, 71–78. [CrossRef]
11. MacLean, A.M.; Bravo, A.; Harrison, M.J. Plant signaling and metabolic pathways enabling arbuscular mycorrhizal symbiosis. *Plant Cell* **2017**, *29*, 2319–2335. [CrossRef]
12. Roth, R.; Paszkowski, U. Plant carbon nourishment of arbuscular mycorrhizal fungi. *Curr. Opin. Plant Biol.* **2017**, *39*, 50–56. [CrossRef]
13. Delaux, P.M. Comparative phylogenomics of symbiotic associations. *New Phytol.* **2017**, *213*, 89–94. [CrossRef] [PubMed]
14. Chen, M.; Bruisson, S.; Bapaume, L.; Darbon, G.; Glauser, G.; Schorderet, M.; Reinhardt, D. VAPYRIN attenuates defence by repressing PR gene induction and localized lignin accumulation during arbuscular mycorrhizal symbiosis of *Petunia hybrida*. *New Phytol.* **2021**, *229*, 3481–3496. [CrossRef] [PubMed]
15. Fernández, I.; Cosme, M.; Stringlis, I.A.; Yu, K.; de Jonge, R.; van Wees, S.M.; Pozo, M.J.; Pieterse, C.M.J.; van der Heijden, M.G.A. Molecular dialogue between arbuscular mycorrhizal fungi and the nonhost plant *Arabidopsis thaliana* switches from initial detection to antagonism. *New Phytol.* **2019**, *223*, 867–881. [CrossRef]
16. Cosme, M.; Fernández, I.; Van der Heijden, M.G.A.; Pieterse, C.M.J. Non-mycorrhizal plants: The exceptions that prove the rule. *Trends Plant Sci.* **2018**, *23*, 577–587. [CrossRef]
17. Cosme, M.; Fernández, I.; Declerck, S.; Van der Heijden, M.G.A.; Pieterse, C.M.J. A coumarin exudation pathway mitigates arbuscular mycorrhizal incompatibility in *Arabidopsis thaliana*. *Plant Mol. Biol.* **2021**, *106*, 319–334. [CrossRef]
18. Hermosa, R.; Botella, L.; Keck, E.; Jiménez, J.Á.; Montero-Barrientos, M.; Arbona, V.; Gómez-Cadenas, A.; Monte, E.; Nicolás, C. The overexpression in *Arabidopsis thaliana* of a *Trichoderma harzianum* gene that modulates glucosidase activity, and enhances tolerance to salt and osmotic stresses. *J. Plant Physiol.* **2011**, *168*, 1295–1302. [CrossRef] [PubMed]
19. Gumz, F.; Krausze, J.; Eisenschmidt, D.; Backenköhler, A.; Barleben, L.; Brandt, W.; Wittstock, U. The crystal structure of the thiocyanate-forming protein from *Thlaspi arvense*, a kelch protein involved in glucosinolate breakdown. *Plant Mol. Biol.* **2015**, *89*, 67–81. [CrossRef] [PubMed]
20. Poveda, J.; Hermosa, R.; Monte, E.; Nicolás, C. *Trichoderma harzianum* favours the access of arbuscular mycorrhizal fungi to non-host Brassicaceae roots and increases plant productivity. *Sci. Rep.* **2019**, *9*, 11650. [CrossRef]
21. Vierheilig, H.; Schweiger, P.; Brundrett, M. An overview of methods for the detection and observation of arbuscular mycorrhizal fungi in roots. *Physiol. Plant.* **2005**, *125*, 393–404. [CrossRef]
22. Lee, J.; Lee, S.; Young, J.P.W. Improved PCR primers for the detection and identification of arbuscular mycorrhizal fungi. *FEMS Microbiol. Ecol.* **2008**, *65*, 339–349. [CrossRef] [PubMed]
23. Yang, H.; Liu, J.; Huang, S.; Guo, T.; Deng, L.; Hua, W. Selection and evaluation of novel reference genes for quantitative reverse transcription PCR (qRT-PCR) based on genome and transcriptome data in *Brassica napus* L. *Gene* **2014**, *538*, 113–122. [CrossRef] [PubMed]
24. Poveda, J.; Hermosa, R.; Monte, E.; Nicolás, C. The *Trichoderma harzianum* Kelch protein ThKEL1 plays a key role in root colonization and the induction of systemic defense in Brassicaceae plants. *Front. Plant Sci.* **2019**, *10*, 1478. [CrossRef] [PubMed]
25. Illescas, M.; Rubio, M.B.; Hernández-Ruiz, V.; Morán-Diez, M.E.; Martínez de Alba, A.E.; Nicolás, C.; Monte, E.; Hermosa, R. Effect of inorganic top dressing and *Trichoderma harzianum* seed-inoculation on crop yield and the shaping of root microbial communities of wheat plants cultivated under high basal N fertilization. *Front. Plant Sci.* **2020**, *11*, 575861. [CrossRef] [PubMed]
26. Livak, K.J.; Schmittgen, T.D. Analysis of relative gene expression data using real-time quantitative PCR and the $2^{-\Delta\Delta CT}$ method. *Methods* **2001**, *25*, 402–408. [CrossRef] [PubMed]
27. De Ollas, C.; González-Guzmán, M.; Pitarch, Z.; Matus, J.T.; Candela, H.; Rambla, J.L.; Granell, A.; Gómez-Cadenas, A.; Arbona, V. Identification of ABA-mediated genetic and metabolic responses to soil flooding in tomato (*Solanum lycopersicum* L. Mill). *Front. Plant Sci.* **2021**, *12*, 613059. [CrossRef] [PubMed]
28. Hart, M.M.; Trevors, J.T. Microbe management: Application of mycorrhyzal fungi in sustainable agriculture. *Front. Ecol. Environ.* **2005**, *3*, 533–539. [CrossRef]
29. Xu, M.J.; Wang, H.Z. Proteomic analysis reveals the mechanisms of *Mycena dendrobii* promoting transplantation survival and growth of tissue culture seedlings of *Dendrobium officinale*. *J. Appl. Microbiol.* **2015**, *118*, 1444–1455. [CrossRef]
30. Sugimura, Y.; Saito, K. Comparative transcriptome analysis between *Solanum lycopersicum* L. and *Lotus japonicus* L. during arbuscular mycorrhizal development. *J. Soil Sci. Plant Nutr.* **2017**, *63*, 127–136. [CrossRef]

31. Takahara, M.; Magori, S.; Soyano, T.; Okamoto, S.; Yoshida, C.; Yano, K.; Sato, S.; Tabata, S.; Yamaguchi, K.; Shigenobu, S.; et al. Too much love, a novel Kelch repeat-containing F-box protein, functions in the long-distance regulation of the legume-*Rhizobium* symbiosis. *Plant Cell Physiol.* **2013**, *54*, 433–447. [CrossRef] [PubMed]
32. Ocampo, J.A. Vesicular-arbuscular mycorrhizal infection of "host" and "non-host" plants: Effect on the growth responses of the plants and competition between them. *Soil Biol. Biochem.* **1986**, *18*, 607–610. [CrossRef]
33. Kabir, Z.; O'Halloran, I.P.; Hamel, C. The proliferation of fungal hyphae in soils supporting mycorrhizal and non-mycorrhizal plants. *Mycorrhiza* **1997**, *6*, 477–480. [CrossRef]
34. Schreiner, R.; Koide, R.T. Antifungal compounds from the roots of mycotrophic and non-mycotrophic plant species. *New Phytol.* **1993**, *123*, 99–105. [CrossRef]
35. Aguilar, R.; Carreón-Abud, Y.; López-Carmona, D.; Larsen, J. Organic fertilizers alter the composition of pathogens and arbuscular mycorrhizal fungi in maize roots. *J. Phytopathol.* **2017**, *165*, 448–454. [CrossRef]
36. Glenn, M.G.; Chew, F.S.; Williams, P.H. Influence of glucosinolate content of *Brassica* (Cruciferae) roots on growth of vesicular–arbuscular mycorrhizal fungi. *New Phytol.* **1998**, *110*, 217–225. [CrossRef]
37. DeMars, B.G.; Boerner, R.E. Arbuscular mycorrhizal development in three crucifers. *Mycorrhiza* **1995**, *5*, 405–408. [CrossRef]
38. Liu, Y.; Xiong, Z.; Wu, W.; Ling, H.Q.; Kong, D. Iron in the symbiosis of plants and microorganisms. *Plants* **2023**, *12*, 1958. [CrossRef] [PubMed]
39. Kabir, A.H.; Debnath, T.; Das, U.; Prity, S.A.; Haque, A.; Rahman, M.M.; Parvez, M.S. Arbuscular mycorrhizal fungi alleviate Fe-deficiency symptoms in sunflower by increasing iron uptake and its availability along with antioxidant defense. *Plant Physiol. Biochem.* **2020**, *150*, 254–262. [CrossRef]
40. Pongrac, P.; Vogel-Mikuš, K.; Poschenrieder, C.; Barcelo, J.; Tolra, R.; Regvar, M. Arbuscular mycorrhiza in glucosinolate-containing plants: The story of the metal hyperaccumulator *Noccaea* (*Thlaspi*) *praecox* (Brassicaceae). In *Molecular Microbial Ecology of the Rhizosphere*; de Brujin, F.J., Ed.; Wiley: Hoboken, NJ, USA, 2013; pp. 1023–1032.
41. Vierheilig, H.; Bennett, R.; Kiddle, G.; Kaldorf, M.; Ludwig-Müller, J. Differences in glucosinolate patterns and arbuscular mycorrhizal status of glucosinolate-containing plant species. *New Phytol.* **2000**, *146*, 343–352. [CrossRef]
42. Cuong, D.M.; Kim, J.K.; Bong, S.J.; Baek, S.A.; Jeon, J.; Park, J.S.; Park, S.U. Comparative analysis of glucosinolates and metabolite profiling of green and red mustard (*Brassica juncea*) hairy roots. *3 Biotech.* **2018**, *8*, 382. [CrossRef] [PubMed]
43. Anthony, M.A.; Celenza, J.L.; Armstrong, A.; Frey, S.D. Indolic glucosinolate pathway provides resistance to mychorrhizal fungal colonization in a non-host Brassicaceae. *Ecosphere* **2020**, *11*, e03100. [CrossRef]
44. Sharma, A.; Sinharoy, S.; Bisht, N.C. The mysterious non-arbuscular mycorrhizal status of Brassicaceae species. *Environ. Microbiol.* **2023**, *25*, 917–930. [CrossRef] [PubMed]
45. Poveda, J.; Díaz-González, S.; Díaz-Urbano, M.; Velasco, P.; Sacristán, S. Fungal endophytes of Brassicaceae: Molecular interactions and crop benefits. *Front. Plant Sci.* **2022**, *13*, 932288. [CrossRef] [PubMed]
46. Tayo, T.; Dutta, N.; Sharma, K. Effect of feeding canola-quality rapeseed mustard meal on animal production—A review. *Agri. Rev.* **2012**, *33*, 114–121.
47. Nour-Eldin, H.H.; Madsen, S.R.; Engelen, S.; Jørgensen, M.E.; Olsen, C.E.; Andersen, J.S.; Seynnaeve, D.; Verhoye, T.; Fulawka, R.; Denolf, P.; et al. Reduction of antinutritional glucosinolates in *Brassica* oilseeds by mutation of genes encoding transporters. *Nat. Biotechnol.* **2017**, *35*, 377–382. [CrossRef] [PubMed]
48. Egert, S.; Baxheinrich, A.; Lee-Barkey, Y.H.; Tschoepe, D.; Stehle, P.; Stratmann, B.; Wahrburg, U. Effects of a hypoenergetic diet rich in α-linolenic acid on fatty acid composition of serum phospholipids in overweight and obese patients with metabolic syndrome. *Nutrition* **2018**, *49*, 74–80. [CrossRef] [PubMed]
49. Scapicchio, P.L. Revisiting choline alphoscerate profile: A new, perspective, role in dementia? *Int. J. Neurosci.* **2013**, *123*, 444–449. [CrossRef]
50. Baum, C.; El-Tohamy, W.; Gruda, N. Increasing the productivity and product quality of vegetable crops using arbuscular mycorrhizal fungi: A review. *Sci. Hortic.* **2015**, *187*, 131–141. [CrossRef]

*agriculture*

MDPI

*Article*

# Preliminary Results of the Impact of Beneficial Soil Microorganisms on Okra Plants and Their Polyphenol Components

**Alaa Abdulkadhim A. Almuslimawi [1,2,†], Lívia László [3,†], Alhassani Leith Sahad [4], Ahmed Ibrahim Alrashid Yousif [1,5], György Turóczi [1] and Katalin Posta [3,6,*]**

[1] Department of Integrated Plant Protection, Plant Protection Institute, Hungarian University of Agriculture and Life Sciences, 2100 Gödöllő, Hungary; almuslimawi.alaa.abdulkadhim.abdulabbass@phd.uni-mate.hu (A.A.A.A.); ahmed.ibrahim.alrashid.yousif.mohamed.elamin@phd.uni-mate.hu (A.I.A.Y.); turoczi.gyorgy@uni-mate.hu (G.T.)

[2] Collage of Agriculture, Al-Qasim Green University, Babylon 51031, Iraq

[3] Department of Microbiology and Applied Biotechnology, Institute of Genetics and Biotechnology, Hungarian University of Agriculture and Life Sciences, 2100 Gödöllő, Hungary; laszlo.livia@uni-mate.hu

[4] Institute of Horticulture Sciences, Hungarian University of Agriculture and Life Sciences, 2100 Gödöllő, Hungary; alhassani.leith@phd.uni-mate.hu

[5] Department of Plant Protection, Omdurman Islamic University, Omdurman 14415, Sudan

[6] Agribiotechnology and Precision Breeding for Food Security National Laboratory, Institute of Genetics and Biotechnology, Hungarian University of Agriculture and Life Sciences, 2100 Gödöllő, Hungary

* Correspondence: posta.katalin@uni-mate.hu

† These authors contributed equally to this work and share first authorship.

**Abstract:** Okra (*Abelmoschus esculentus* L.) is a highly nutritious vegetable rich in vitamins, minerals, and bioactive compounds, including polyphenols, offering numerous health benefits. Despite its nutritional value, okra remains underutilized in Europe; however, its cultivation and popularity may rise in the future with increasing awareness of its advantages. In agricultural practices, beneficial soil microorganisms, such as arbuscular mycorrhizal fungi (AMF), *Trichoderma* spp., *Streptomyces* spp., and *Aureobasidium* spp., play crucial roles in promoting plant health, enhancing agricultural productivity together with improved crop nutritional value. This study aimed to investigate the effects of individual and combined inoculation on the polyphenol content of okra fruits, as analyzed by HPLC. Moreover, growth parameters and glutathione-S-transferase enzyme (GST) activities of okra leaves were also estimated. Tested microorganisms significantly increased the yield of okra plants except for *A. pullulans* strain DSM 14950 applied individually. All microorganisms led to increased GST enzyme activity of leaves, suggesting a general response to biotic impacts, with individual inoculation showing higher enzyme activity globally compared to combined treatments. According to the polyphenol compound analysis, the application of tested microorganisms held various but generally positive effects on it. Only the combined treatment of *F. mosseae* and *Streptomyces* strain K61 significantly increased the coumaric acid content, and the application of *Aureobasidium* strain DSM 14950 had a positive influence on the levels of quercetin and quercetin-3-diglucoside. Our preliminary results show how distinct polyphenolic compound contents can be selectively altered via precise inoculation with different beneficial microorganisms.

**Keywords:** okra; *Funneliformis mosseae*; *Trichoderma* spp.; *Aureobasidium* spp.; *Streptomyces* spp.; polyphenol

## 1. Introduction

Okra (*Abelmoschus esculentus* L.) is a significant member of the Malvaceae family. It originated in Ethiopia; today, it is still a widely cultivated and consumed vegetable plant in tropical and sub-tropical regions [1]. Although the production of okra in Europe is not significant, it has increased in recent years, reaching 8094.37 t in 2022. Because of its origin, okra is frost-sensitive and vulnerable to drought conditions, and it thrives in regions characterized by warm climates and ample sunlight [2–4]. Its optimal germination temperature ranges between 28 and 32 °C; therefore, some warmer areas of Europe are also suitable for okra cultivation.

Okra pods have gastronomical and health benefits related to their rich nutrient content and essential mineral content [5–8]. They abound in microelements, like Fe, Zn, Mn, and Ni [9]; moreover, they are also rich in various vitamins, such as vitamins C, B, E, and K [10]. Both vitamins C and E maintain antioxidant functions and play a vital role in reducing and controlling oxidative stress [11,12]. In addition, okra fruits are also rich in polyphenols [13], which represent a diverse group of naturally occurring compounds found abundantly in plants, characterized by the presence of multiple phenolic rings, and are known for their antioxidant properties [14,15]. Polyphenols have been extensively studied for their potential health benefits, including anti-inflammatory, anticancer, antimicrobial, and cardiovascular protective effects. Their antioxidant attribution and their abundance establish them as one of the major bioactive compounds of okra fruits [16–19].

Okra plants are not widely known and consumed in Europe; however, due to the beneficial properties of this vegetable previously described, its popularity and relevant outcomes will increase in the future. This opportunity requires the gathering of information about the cultivation of okra plants, and considering global climatic trends, it is pivotal to broaden the knowledge about the application of sustainable methods. Among these technologies, there is a popular method of using beneficial soil microorganisms which play crucial roles in promoting plant health and enhancing agricultural productivity. These microorganisms have an impact on plants through various mechanisms, both direct and indirect ways to influence nutrient acquisition, disease suppression, and stress tolerance. Among the beneficial fungi, arbuscular mycorrhizal fungi (AMF) represent a major part of plant–microbe interactions according to their abundance and great beneficial impacts [20]. Furthermore, they also improve the secondary metabolite production and nutritional values of different crops [21]. Another beneficial fungus, *Trichoderma asperellum*, generally occurs in soil, as does *Streptomeyces griseoviridis*, a Gram-positive bacterium; besides their antagonistic trait, their saprophytic nature is also well-known. Both these genera are characterized by the production of secondary metabolites, which provides opportunities for practical applications [22,23]. In addition to these biocontrol microorganisms, *Aureobasidium pullulans* has also emerged as a new alternative, further expanding the existing possibilities [24].

In our study, we inquire if these microorganisms' combined application with AMF shows possible synergistic combinations despite the fact that a major obstacle to beneficial impacts may be their antagonistic properties. Moreover, we aimed to gain deeper insight into how different inoculants influence the quantity of polyphenol components.

## 2. Materials and Methods

### 2.1. Plant Material and Maintaining the Strains

The commercial seeds of okra (*Abelmoschus esculentus* L. 'Moench' var. Lady Finger F1) originated from a company (Agrimax group S.L.U, Barcelona, Spain).

Arbuscular mycorrhizal (AM) fungi, based on spore morphology and molecular methods, were identified as *Funneliformis mosseae* (Glomerales, Glomeraceae). This strain originated from the collection of the Department of Microbiology and Applied Biotechnology of the Hungarian University of Agriculture and Life Sciences, Hungary. *F. mosseae* was propagated on maize (*Zea mays* L. 'Golda F1') growing on sterilized peat (Klasmann TS3, 100 mg $L^{-1}$ $P_2O_5$) and sand 1:3 ($v/v$) substrate for three successive propagation cycles, each lasting 5 months. The most probable number (MPN) of infective propagules (approximately 35 infective propagules $g^{-1}$) was determined following the method of Feldmann and Idczack [25].

*Trichoderma asperellum* strain T34 of a filamentous fungus, a commercial product named Xilon (Kwizda Agro GmbH, Wien, Austria), was selected for this research. It was cultured on Potato Dextrose Agar medium (Difco PDA, 20 g/L dextrose, 15 g/L agar, and 4 g/L potato starch) at 25 °C (ambient temperature and illumination) for a week.

*Aureobasidium pullulans* strain DSM 14950 originated from Blossom Protect (SAN Agrow Holding GmbH, Herzogenburg, Austria), a commercial product. It was cultured on Potato Dextrose Agar medium (Difco PDA, 20 g/L dextrose, 15 g/L agar, and 4 g/L potato starch).

*Streptomyces* strain K61, previously *S. griseoviridis*, was from Lalstop K61 WP (formerly Mycostop, Danstar Ferment AG, Zug, Switzerland), which is also a commercial product. We used dextrose, 15 g agar, and 4 g potato starch in 1 L of distilled and bacterial media, Nutrient Agar (NA), 5 g of peptic digest of animal tissue, 3 g of beef extract, and 15 g of agar in 1 L of distilled media to subculture it.

### 2.2. Experimental Design

The experiments were conducted during the summer of 2022 in the experimental station belonging to the Plant Protection Institute—Hungarian University of Agriculture and Life Sciences (MATE), located at 2100 Gödöllő, Pest, Hungary (coordinates: 47.594315, 19.368984).

The okra plants were initially cultivated in a greenhouse environment, which is renowned for maintaining temperature conditions typically ranging between 18 and 35 °C, and the duration of illumination was 12–14 h of light per day in a greenhouse. The applied soil type in this experiment was sterilized sand mixed with horticultural soil (Klasmann TS3, 100 mg $L^{-1}$ $P_2O_5$) 3:1 ($v/v$); the juvenile plants were cultivated in small pots, and subsequently, they were transferred to bigger pots containing 2 kg of sterilized soil mixture with one plant in each pot. Soil pH was 6.7. Pots were placed in glasshouse benches in a completely randomized design for each treatment. The plants were subjected to a watering regimen, with irrigation occurring every two days or as needed, by monitoring the moisture content of the soil.

Plants were treated with three microorganism strains and AMF individually and in combination. The experiment setup contained 11 replicates of okra plants per treatment.

AMF inoculum was introduced to okra plants through an inoculation process involving the addition of 10 g mycorrhizal inoculum containing approximately 350 infective propagules per plant. The control plants received the same amount of inoculum after it was sterilized three times.

Spore suspensions ($4.5 \times 10^6$ spores $mL^{-1}$, $8.8 \times 10^6$ spores $mL^{-1}$, and $1.1 \times 10^6$ spores $mL^{-1}$, respectively) of *T. asperellum* strain (T34), *A. pullulans* strain (DSM 14950), and *S. griseoviridis* (K61) were prepared after cultivation on PDA medium or Nutrient Agar at 25 °C for 7 days. Plants were irrigated with the respective suspensions at 20 D when they had at least nine or more unfolded leaves on the main shoot.

Plants were harvested after 75 days.

### 2.3. Determination of the Root Colonization by Arbuscular Mycorrhizal Fungi

After the 75-day-old plants were harvested, root samples of 5 plants per treatment were randomly taken to identify mycorrhiza colonization of the root. Root samples of okra plants were washed thoroughly with tap water and then cut into 1 cm long pieces. The roots were placed in a 10% aqueous solution of KOH ($w/v$). Roots and KOH were heated in a water bath at 90 °C for 60 min; then, the solution was decanted, and the roots were washed with running water; after that, the roots were washed with 5% vinegar (acetic acid) for 1–2 min and put in 5% ink (Pelikan blue), and then the roots and the ink were boiled for 2 min [26]. Root colonization was evaluated through visual observation using a stereomicroscope with a magnification of 100×. To determine the colonization, the gridline intersection method was used [27]. This involved observing the presence or absence of mycorrhizal structures at the intersections between the root fragments and the gridlines.

### 2.4. Assessment of the Growth Parameters

The shoot and root fresh weight (FW) of each sample were weighed after harvesting (75D), and then each sample was dried in a hot-air oven at 70 °C for 2 days to determine its dry weight (DW).

The yield of okra can be measured in terms of the quantity of pods harvested per treatment. Fruit was collected from the plants between the 54th and 75th days.

### 2.5. Determination of Glutathione-S-Transferase Enzyme Activity

During the sampling process (D60) for measuring glutathione-S-transferase (GST) enzyme activity, plant leaves (the third leaf from the top of the plant, which is healthy) were collected and stored at −80 °C until utilization.

Leaf tissue (500 mg) was suspended in 100 µL of Cell Lysis Buffer; then, leaves were homogenized with a mortar and pestle. Samples were centrifuged for 10 min at 4 °C and 13,300 rpm using a cold microcentrifuge to remove any insoluble material. The supernatant was transferred into a clean tube and kept on ice. GST activity was measured based on the method of Habig et al. [28]. The soluble protein level of all extracts was determined according to the method of Bradford [29].

### 2.6. HPLC Analysis of the Polyphenol Compounds

For the purpose of conducting phenol analysis, the fruit of the okra plants was physically harvested by handpicking every two to three days, starting from the commencement of the flowering and fruiting period. It was then preserved in a biotechnology laboratory of MATE at a temperature of −20 degrees Celsius.

Then, 300 mg of lyophilized (freeze-dried) whole okra fruits was taken and crushed in a crucible mortar in the presence of 1–2 g of quartz sand. The phenolic compounds were extracted by adding methanol containing 2% orthophosphoric acid. The macerate was then transferred to a centrifuge tube and subjected to ultrasonication for 15 min at 40 °C in a water-bath ultrasonic device (model RK-165-BH Bendelin Sonorex, Berlin, Germany) followed by mechanical shaking at room temperature for 20 min. The extract was centrifuged for 5 min at 5000 rpm (M-Universal, MPW Med. Instrument, Warsaw, Poland). The supernatant was decantated into a round-bottom flask, and the solvent was evaporated to dryness under vacuum at 45 °C. The residues were redissolved in 5 mL of 1:1 methanol/L% orthophosphoric acid and finally purified through a 0.45 µm, 25 mm cellulose acetate syringe filter before injection into the HPLC apparatus.

A Hitachi Chromaster HPLC instrument (Tokyo, Japan) containing a Hitachi Chromaster Model 5160 gradient pump (Tokyo, Japan), a Hitachi Chromaster Model 5260 autosampler, a Hitachi Chromaster Model 5310 column oven (Tokyo, Japan), and a Hitachi Chromaster Model 5430 diode-array detector (Tokyo, Japan) was used with Agilent OpenLab EZChrom A.04.10 software (Santa Clara, CA, USA) for operation and data processing.

The separation of phenolic compounds was performed on an Ascentis phosphor-conditioned C18 phase (C18-PCP, from Supelco, Bellefonte, PA, USA) with gradient elution of 1% ortho-phosphoric acid (A) and acetonitrile (B) according to a recently developed protocol (under publication). The gradient elution started with 1% B in A, changed to 20% B in 20 min, stayed isocratic for 10 min, changed to 30% B in 5 min, stayed isocratic for 10 min, and finally turned to 1% B in 5 min. The DAD detection was between 190 nm and 700 nm. The quantification was based on recording the area at the maximum absorbance wavelength of each compound and relating it to that of the standard solution.

Stock solutions for different phenolics (Sigma-Aldrich via Merck, Budapest, Hungary) were prepared by dissolving 2–3 mg in 10 mL absolute ethanol or methanol and diluted 10 times with 40% ethanol in 1% ortho-phosphoric acid. The working solutions were used for calibration curves, identification, and quantification of phenolic compounds; in case no standard was available, the compounds were tentatively identified based on a comparison of their spectral characteristics and chromatographic behavior with literature data [30–32].

### 2.7. Statistical Analysis

RStudio 2024.04.0+735 software was used for statistical analysis. All data were evaluated by one-way analysis of variance (ANOVA) with various microorganism applications. Means were compared by Tukey post hoc test at $p < 0.05$, and Pearson correlation coefficients (R) among physiological parameters, GST enzyme activity, and polyphenol contents of okra fruits were calculated. In addition, PCA was carried out to identify patterns, i.e., interactions among the studied variables and treatments, in the polyphenolic data of okra fruits treated with different beneficial microorganisms and their combinations. The results of the statistical analysis are presented in Tables A1–A5 in the Appendix A.

## 3. Results

### 3.1. AMF Colonization

The plants inoculated with arbuscular mycorrhizal fungi showed an appropriate level of colonization (Figure 1); applying *T. asperellum* strain T34 (M + T) or *A. pullulans* strain DSM 14950 (M + A) slightly decreased the colonization level; however, the application of *S. griseoviridis* strain K61 (M + S) did not influence mycorrhizal presence. The highest colonization level was reached by using AMF individually (M) and AMF combined with *S. griseoviridis* strain K61 and was 84.68% and 85.16%, respectively.

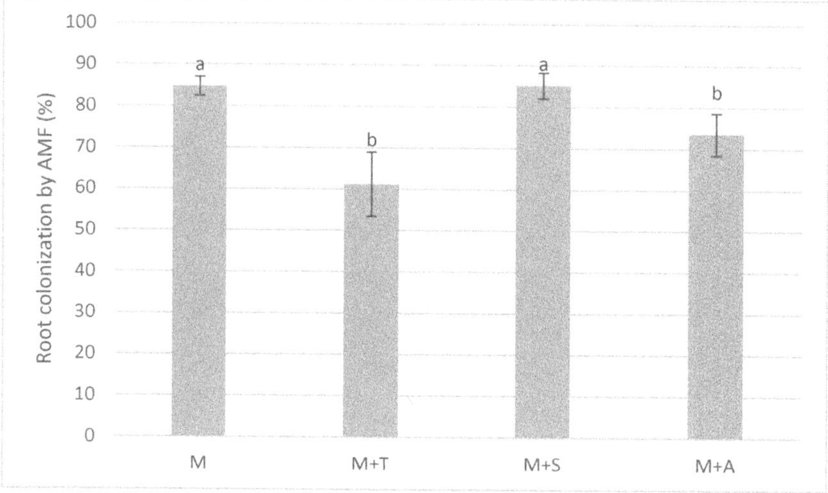


**Figure 1.** Root colonization level of the okra plants inoculated with arbuscular mycorrhizal fungi. M: *F. mosseae*; M + T: *F. mosseae* and *T. asperellum* strain T34; M + S: *F. mosseae* and *S. griseoviridis* strain K61; M + A: *F. mosseae* and *A. pullulans* strain DSM 14950. Mean values followed by the same alphabets did not differ significantly based on Tukey's test ($p < 0.05$).

### 3.2. Fresh and Dry Weight of Plant and Yield Quality

According to our results, the treatments in which we applied AMF increased the plants' fresh and dry weight (Figure 2). When we applied only *T. asperellum* strain T34 (T), *S. griseoviridis* strain K61 (S), or *A. pullulans* strain DSM 14950 (A), the treatment did not have a significant impact on the weights compared to the control plants (C). In the case of the combined treatments, the beneficial effect of the AMF prevailed; when it was applied with *T. asperellum* strain T34 (M + T) or *A. pullulans* strain DSM 14950 (M + A), the measured shoot and root weights were the same as when we applied AMF individually (M); therefore, we did not observe the antagonist effect of the other fungi. When we applied AMF and *S. griseoviridis* strain K61 (M + S) in combined treatment, we recorded lower weight data compared to the plants that were inoculated only with AMF.

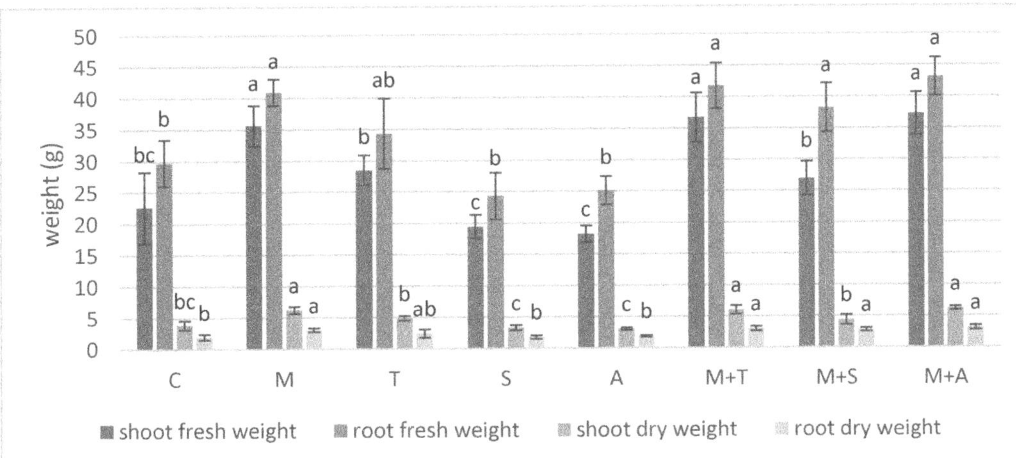


**Figure 2.** Average fresh and dry weight of the okra plants' shoots and roots. C: no treatment; M: *F. mosseae*; T: *T. asperellum* strain T34; S: *S. griseoviridis* strain K61; A: *A. pullulans* strain DSM 14950; M + T: *F. mosseae* and *T. asperellum* strain T34; M + S: *F. mosseae* and *S. griseoviridis* strain K61; M + A: *F. mosseae* and *A. pullulans* strain DSM 14950. Mean values followed by the same alphabets did not differ significantly based on Tukey's test ($p < 0.05$).

According to our results, the yield of the okra plants is significantly elevated by the applied microorganisms (Figure 3), except for the treatment in which we applied *A. pullulans* strain DSM 14950 individually (A). This yeast-like fungus also had a negative effect on the beneficial impact of the AMF; in the combined treatment (M + A), we did not record as high a yield as the other cases.

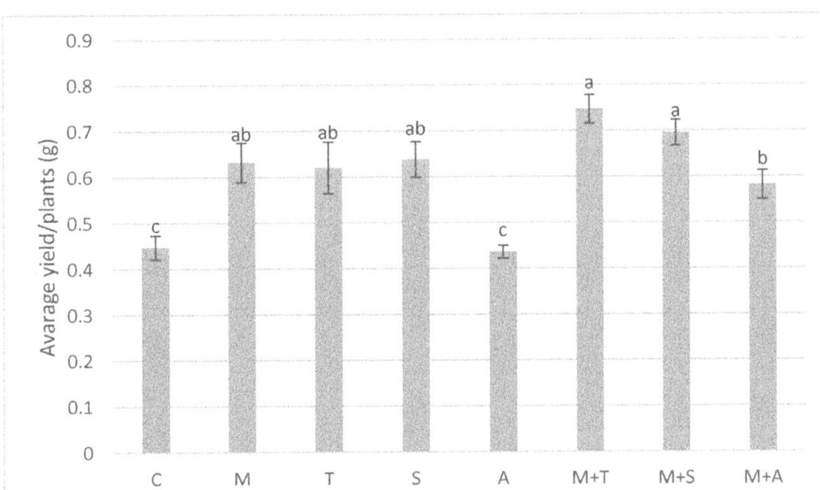


**Figure 3.** The average yield of the okra fruits/plants. C: no treatment; M: *F. mosseae*; T: *T. asperellum* strain T34; S: *S. griseoviridis* strain K61; A: *A. pullulans* strain DSM 14950; M + T: *F. mosseae* and *T. asperellum* strain T34; M + S: *F. mosseae* and *S. griseoviridis* strain K61; M + A: *F. mosseae* and *A. pullulans* strain DSM 14950. Mean values followed by the same alphabets did not differ significantly based on Tukey's test ($p < 0.05$).

### 3.3. GST Enzyme Activity

We observed elevated levels of GST enzyme activity when we applied beneficial microorganisms (Figure 4). Combined treatments also increased GST enzyme activity, although when we applied AMF and *S. griseoviridis* strain K61 together, they did increase the GST enzyme activity compared to the control plants (C) but not as much as other treatments. Further studies are needed to broaden the knowledge about the interaction between the applied microorganisms and their effect on the GST enzyme activity.

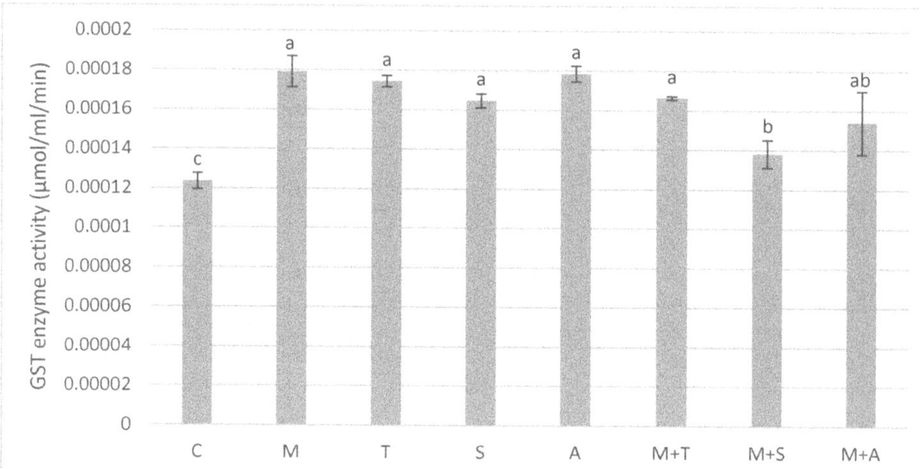


**Figure 4.** GST enzyme activity in the leaf of the okra plants. C: no treatment; M: *F. mosseae*; T: *T. asperellum* strain T34; S: *S. griseoviridis* strain K61; A: *A. pullulans* strain DSM 14950; M + T: *F. mosseae* and *T. asperellum* strain T34; M + S: *F. mosseae* and *S. griseoviridis* strain K61; M + A: *F. mosseae* and *A. pullulans* strain DSM 14950. Mean values followed by the same alphabets did not differ significantly based on Tukey's test ($p < 0.05$).

### 3.4. Effects of Biotic Treatments on Phenolic Compounds of Fruit

One of the important objectives of the present work is to investigate the effect of mycorrhiza alone and in combination with microbes on the content of polyphenols in okra fruits. The HPLC protocol applied to analyze polyphenols from the whole okra fruit allowed for excellent separation of the main compounds and their derivatives, mainly dimers and glycosides (Figure 5). To make the discussion easier and more meaningful, the obtained results are arranged in groups for the main compounds.

Figure 6 shows the effect of mycorrhiza and antagonistic microorganisms on the content of kaempferol derivatives in the whole okra fruit. The combination of arbuscular mycorrhizal fungi with other microorganisms in different treatments (M, T, S, M + T, M + S, M + A) caused a significant increase in the level of okra fruit phenols as compared to the control treatment (C). The results presented in Figure 6 demonstrate the effects of various treatments on kaempferol content in okra fruit, in conjunction with arbuscular mycorrhizal fungi and other microorganisms. Figure 6 shows the impact of the combination of mycorrhiza and microbe strains on the okra in increasing the percentage of fruit phenols (kaempferol derivatives). The inoculation of the plants with *T. asperellum* strain T34 (T), mycorrhiza fungi (M), *Streptomyces* strain K61 (S), M + S, M + T, M + A, and *A. pullulans* strain DSM 14950 (A) increased the content of kaempferol by 38%, 29%, 27%, 22%, 20%, 17%, and 14%, respectively, compared with the control, and these increases were highly significant. The highest increase in the content of the kaempferol derivatives was recorded for T treatment.

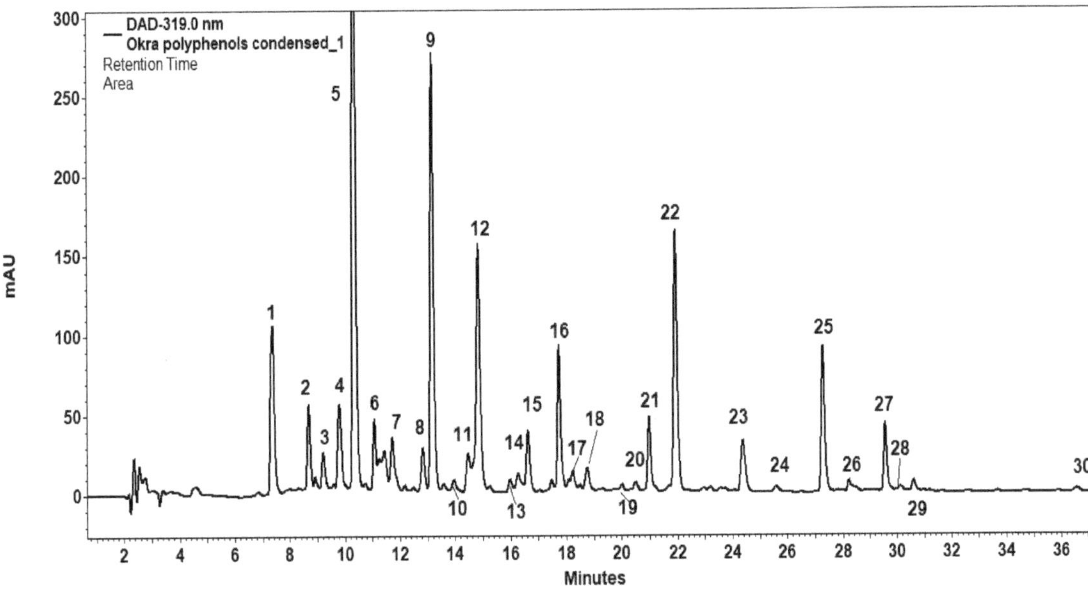


**Figure 5.** HPLC profile of phenolic compounds extracted from lyophilized okra fruits and separated on C18-PCP column with gradient elution of acetonitrile in 1% orthophosphoric acid. Detection was at 319 nm. Peak identification is shown in Table A6 in the Appendix A.

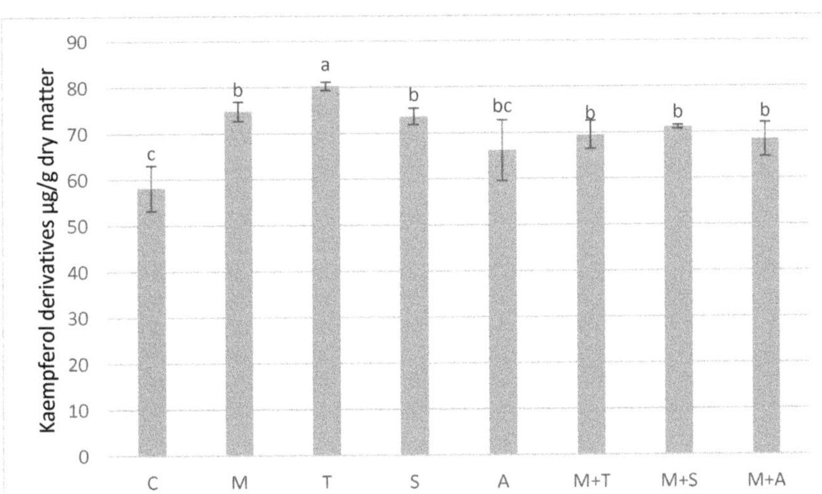


**Figure 6.** Effect of AMF and antagonistic microorganisms on kaempferol derivatives in okra fruit. C: no treatment; M: *F. mosseae*; T: *T. asperellum* strain T34; S: *S. griseoviridis* strain K61; A: *A. pullulans* strain DSM 14950; M + T: *F. mosseae* and *T. asperellum* strain T34; M + S: *F. mosseae* and *S. griseoviridis* strain K61; M + A: *F. mosseae* and *A. pullulans* strain DSM 14950. Mean values followed by the same alphabets did not differ significantly based on Tukey's test ($p < 0.05$).

The results presented in Figure 7 demonstrate the response of coumaric acid derivatives to various treatments. Such important polyphenolic compounds exist in okra fruits in a free form and as coumaroyl-hexoside. The treatments of M + S and M + A showed an increase in the content of the two derivatives as compared to the control group. It is evident that a highly significant increase of coumaric acid and coumaroyl hexoside was recorded for treatments of M + A and M + S, respectively. As compared to the control, other treatments either decreased or had no significant effect on the level of coumaric acid derivatives. The results showed that the plants' inoculation with microbes increased the level of coumaroyl derivatives by 43%, 22%, 12%, 9%, and 7% with M + A, M + S, M, S, and A, respectively. As for *T. asperellum* strain T34, no significant effect on coumaric acid in the okra plant was noticed.

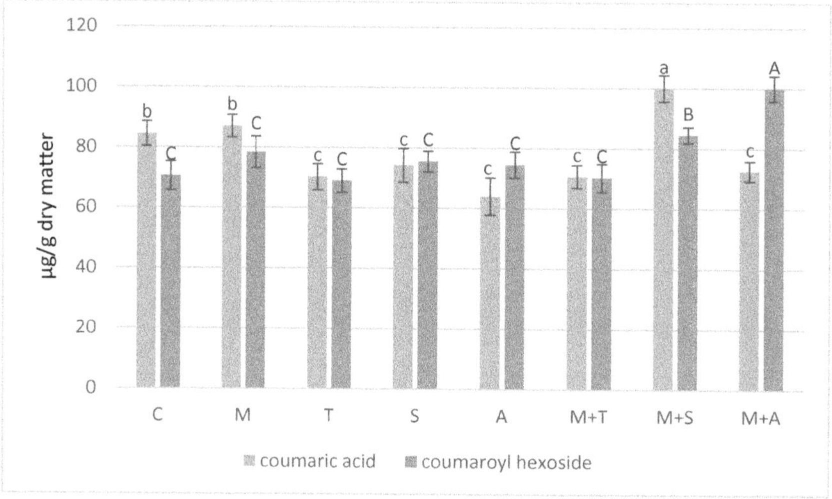


**Figure 7.** Effect of AMF and antagonistic microorganisms on coumaric acid and coumaroyl hexoside content in okra fruit. C: no treatment; M: *F. mosseae;* T: *T. asperellum* strain T34; S: *S. griseoviridis* strain K61; A: *A. pullulans* strain DSM 14950; M + T: *F. mosseae* and *T. asperellum* strain T34; M + S: *F. mosseae* and *S. griseoviridis* strain K61; M + A: *F. mosseae* and *A. pullulans* strain DSM 14950. Mean values followed by the same alphabets did not differ significantly based on Tukey's test ($p < 0.05$).

Quercetin was found to exist in the extract of the whole okra fruit as quercetin-3-o-glucoside, quercetin-di-glucoside, and quercetin-3-O-(melanoyl)glucoside. The impact of the different treatments on the quantity of quercetin derivatives is shown in Figure 8. It is of interest that *A. pullulans* strain DSM 14950 (A) treatment resulted in a highly significant increase in the average content of quercetin, which was found to be due to higher activation of quercetin-3-diglucoside biosynthesis. Interestingly, the combination of mycorrhiza with *A. pullulans* strain DSM 14950 (M + A) yielded fruits with significantly higher levels of quercetin 3-diglucoside as compared to other treatments, but not as high as those recorded for *A. pullulans* strain DSM 14950 alone.

Distinct patterns were found in the accumulation of chlorogenic acid–catechin–glucoside phenols in okra fruit under the influence of arbuscular mycorrhizal fungi and antagonistic microorganisms M, T, A, and S and the combination of AMF with other microorganisms: M + T, M + A, and M + S. Quantitative measurements allowed for the characterization of phenolic content, with specific attention to variations induced by the microbial treatments. The biotic treatments with M, M + T, M + S, Sg, M + A, T, and A increased the level of these polyphenols by 163%, 116%, 86%, 79%, 63%, 29%, and 5%, respectively, and it was found that M and M + T treatments significantly increased the content of chlorogenic acid–catechin dimers in okra fruits as compared to the control (Figure 9).

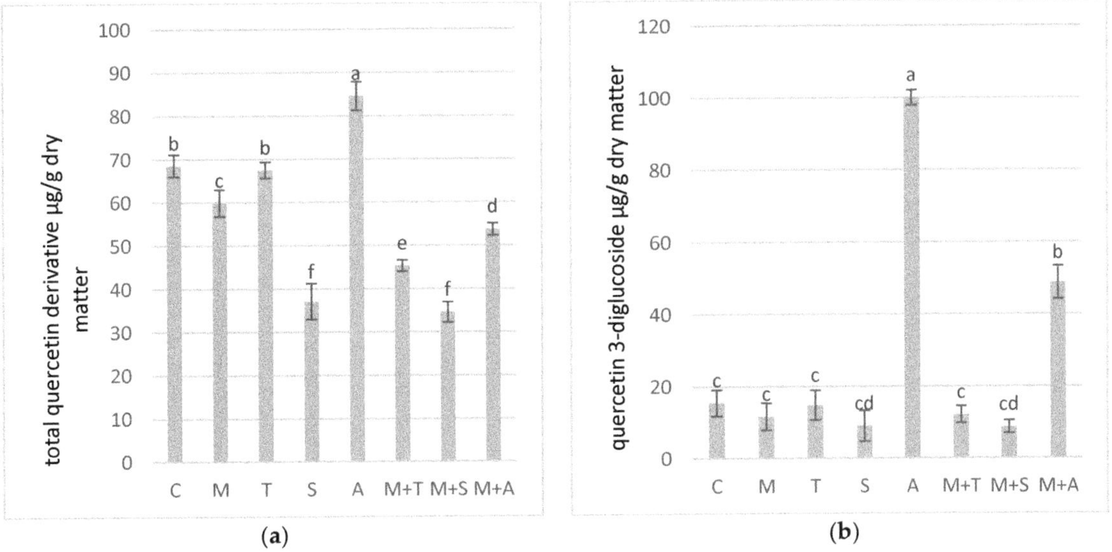


**Figure 8.** Impact of different biotic treatments on the content of total quercetin derivatives (**a**) and quercetin 3-diglucoside (**b**) in okra fruits. C: no treatment; M: *F. mosseae;* T: *T. asperellum* strain T34; S: *S. griseoviridis* strain K61; A: *A. pullulans* strain DSM 14950; M + T: *F. mosseae* and *T. asperellum* strain T34; M + S: *F. mosseae* and *S. griseoviridis* strain K61; M + A: *F. mosseae* and *A. pullulans* strain DSM 14950. Mean values followed by the same alphabets did not differ significantly based on Tukey's test ($p < 0.05$).

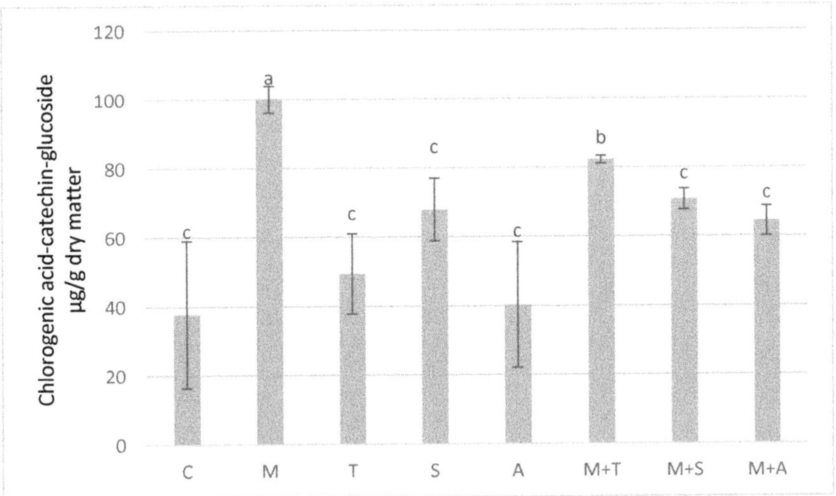


**Figure 9.** Effect of AMF and antagonistic microorganisms on chlorogenic acid–catechin–glucoside in okra fruit. C: no treatment; M: *F. mosseae*; T: *T. asperellum* strain T34; S: *S. griseoviridis* strain K61; A: *A. pullulans* strain DSM 14950; M + T: *F. mosseae* and *T. asperellum* strain T34; M + S: *F. mosseae* and *S. griseoviridis* strain K61; M + A: *F. mosseae* and *A. pullulans* strain DSM 14950. Mean values followed by the same alphabets did not differ significantly based on Tukey's test ($p < 0.05$).

It is evident from the results presented in Figure 10 that all treatments of mycorrhiza alone or combined with other microbial strains, except for the combination of AMF with *S. griseoviridis* strain K61 (M + S), promoted the biosynthetic pathways of sinapic acid derivatives as compared to the control treatment. The highest increase in the content of such dimers was found in fruits of okra treated with AMF combined with *A. pullulans* strain DSM 14950.

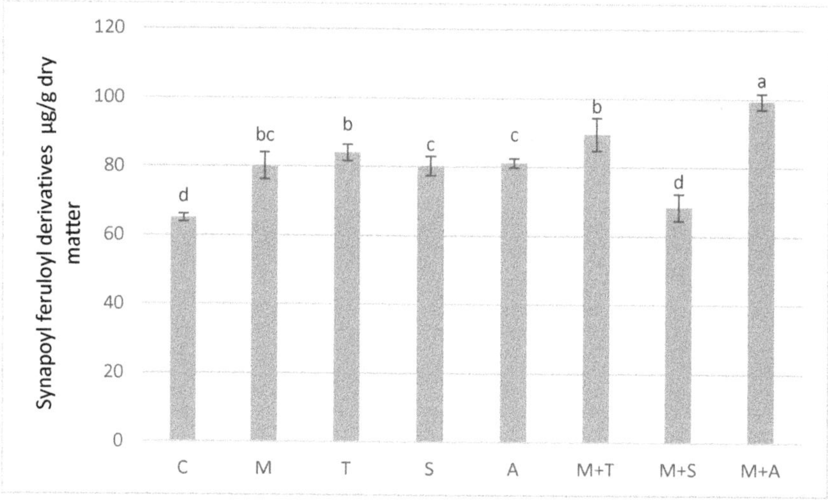


**Figure 10.** Effect of AMF and antagonistic microorganisms on total content of sinapoyl feruloyl derivatives in okra fruit. C: no treatment; M: *F. mosseae;* T: *T. asperellum* strain T34; S: *S. griseoviridis* strain K61; A: *A. pullulans* strain DSM 14950; M + T: *F. mosseae* and *T. asperellum* strain T34; M + S: *F. mosseae* and *S. griseoviridis* strain K61; M + A: *F. mosseae* and *A. pullulans* strain DSM 14950. Mean values followed by the same alphabets did not differ significantly based on Tukey's test ($p < 0.05$).

In Figure 11, the synergistic impact of the combined application of mycorrhiza and other microorganisms on di-caffeoylquinic acid levels in okra fruits is depicted. All treatments except S and M + S treatments showed significantly higher contents of di-caffeoylquinic acid in okra fruits. The highest increase in the concentration of this polyphenol was found in M, M + T, and M + A with no significant variation between them in their impact on di-caffeoylquinic acid. In particular, the S-type bacterial inoculation had a remarkable negative impact on the metabolic pathway of this polyphenol, most probably due to the partial inhibition of the enzymes involved in the biosynthesis processes of di-caffeoylquinic acid. The combination of S with AMF slightly moderated the negative effect of S on di-caffeoylquinic acid formation in okra fruit; however, the difference was not significant.

Principal component analysis of individual polyphenols was performed, and its result is shown in Figure 12. The first two principal components (PC1 and PC2) explain 59.7% of the total variation. Three components showed eigenvalues higher than 1 (Table A4 in the Appendix A). PC1, covering 33.5% of the total variation, had negative associations with quercetin and quercetin-3-diglucoside and positive associations with coumaric acid and chlorogenic acid. In addition, 26.2% of the total variance was covered by PC2, which was positively influenced by kaempferol, coumaroyl hexoside, chlorogenic acid, sinapoyl, and di-caffeylquinic acid (Table A5 in the Appendix A).

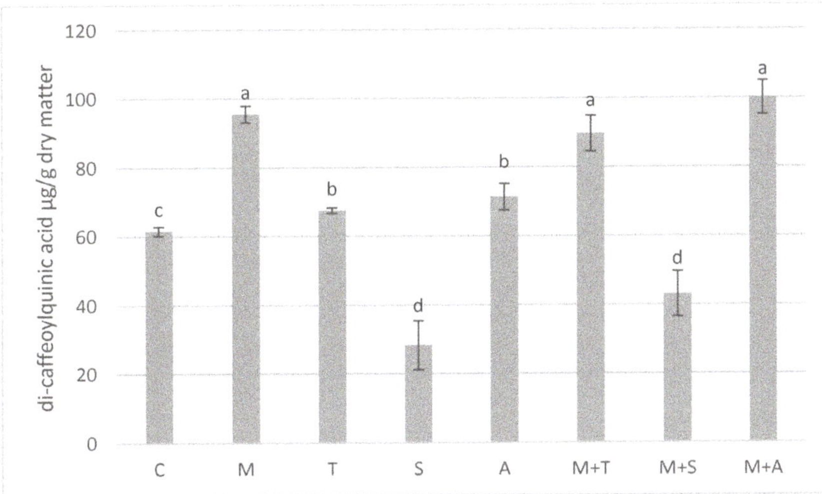


**Figure 11.** Effect of AMF and antagonistic microorganisms on di-caffeoylquinic acid in okra fruit. C: no treatment; M: *F. mosseae*; T: *T. asperellum* strain T34; S: *S. griseoviridis* strain K61; A: *A. pullulans* strain DSM 14950; M + T: *F. mosseae* and *T. asperellum* strain T34; M + S: *F. mosseae* and *S. griseoviridis* strain K61; M + A: *F. mosseae* and *A. pullulans* strain DSM 14950. Mean values followed by the same alphabets did not differ significantly based on Tukey's test ($p < 0.05$).

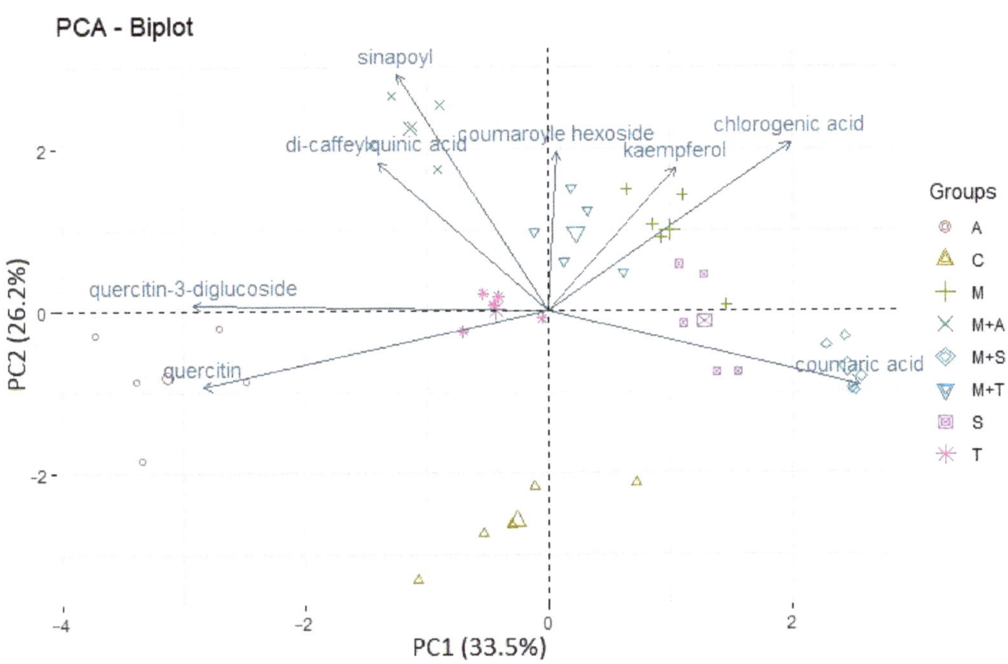


**Figure 12.** Principal component analysis of individual polyphenols in the fruits of okra plants. C: no treatment; M: *F. mosseae*; T: *T. asperellum* strain T34; S: *S. griseoviridis* strain K61; A: *A. pullulans* strain DSM 14950; M + T: *F. mosseae* and *T. asperellum* strain T34; M + S: *F. mosseae* and *S. griseoviridis* strain K61; M + A: *F. mosseae* and *A. pullulans* strain DSM 14950.

The biplot demonstrated relatively clear discrimination among the groups of the control treatment and the treatments with beneficial microorganisms. Differences among A, C, and M + S groups were discriminated by PC1, while PC2 distinguished between the beneficial microbes and the control treatment.

Coumaric acid content showed a strong negative correlation with the sinapoyl, quercetin-3-diglucoside, and quercetin contents of the okra fruits according to the Pearson correlation analysis (Table A3 in the Appendix A). There were positive correlations between kaempferol and chlorogenic acid, as well as between sinapoyl and coumaroyl hexoside.

Some of the polyphenolic compounds of the okra fruit showed a strong correlation with the yield (Table A3 in the Appendix A); kaempferol and chlorogenic acid had a positive association; meanwhile, quercetin-3-diglucoside and quercetin had a negative connection with it. The kaempferol and the sinapoyl content showed a positive correlation with the GST enzyme activity, while in the case of coumaric acid content, their connection was described negatively. Obviously, strong positive relationships were also found between all the growth parameters.

## 4. Discussion

Okra is a highly nutritious vegetable due to its rich vitamin and mineral content. Furthermore, the presence of bioactive compounds, including polyphenols, offers numerous advantageous impacts on human health. However, despite its nutritional value and potential benefits, okra is not widely consumed in Europe, but its cultivation and popularity may increase in the future with growing awareness of its opportunities; therefore, it is pivotal to gather more scientific results about how cultivation methods influence its contents of beneficial secondary metabolites. Vegetable plants like okra are exposed to several biotic and abiotic factors that influence their secondary metabolic pathways and thereby alter the quantity and quality of their bioactive compounds [33]. In our work, well-known beneficial microorganisms like *F. mosseae*, *T. asperellum* strain T34, *S. griseoviridis* strain K61, and *A. pullulans* strain DSM 14950 and their combinations were applied to test their influence on okra's growth and polyphenol content.

Among the applied microorganisms, *F. mosseae* and *T. asperellum* strain T34 showed the best plant-growth-promoting effect, similar to the work of Ali et al. and Mwangi et al. [34,35]. These tendencies were manifested both in individual and combined treatments. Interestingly, mixed inoculation always caused higher growth responses than separate inoculation, and a synergistic relationship between inoculants could be observed, but not for M + T. Some references also highlight that microorganisms living in the rhizosphere could influence the expression of plant–AMF symbiosis [36–38]. Estimating the root colonization together with the growth parameters of okra, our results confirmed that the antagonism of *Trichoderma* strain T34 against AMF could be compensated for by the saprophytic ability of the mentioned fungi. The degradation of organic matter in the soil and the release of nutrients by *Trichoderma* are well-known [39,40], and the bridge offered by the external mycelium of AMF between the plant and nutrient-rich patches provides support for the plant. In this way, even the putative negative effects of biocontrol strains, like *Trichoderma* spp., can be compensated for.

Different classes of polyphenols, including flavonoids (e.g., flavonols, flavanols, flavones, flavanones, anthocyanins), phenolic acids (e.g., hydroxybenzoic acids, hydroxycinnamic acids), and other polyphenols (e.g., lignans, stilbenes), are important components of okra fruits [41]. Although there are some studies about the polyphenol content of okra fruits inoculated with or without AMF [42–44], our work is novel in the scope of changes in various polyphenols in the fruit of okra under the influences of different microorganisms.

Our preliminary results, based on PCA, clearly showed that the estimated polyphenol profile can be altered by the applied microorganisms. Coumaric acid, a well-known antioxidant, increased significantly only with the combined treatment of *F. mosseae* and *S. griseoviridis* strain K61. This finding is contradictory to other studies that reported a positive effect of AMF on coumaric acid [45,46], where different plant species, like cucumber (*Cucumis sativus* L.) or strawberry (*Fragaria* × *ananassa* Duch.), and AMF were used. This result highlights that different plant and AMF species isolates can modulate plant metabolites on distinct levels.

Another important polyphenol is quercetin, which has antioxidant characteristics and is also influenced by AMF [47]. Eftekhari et al. presented a generally positive effect on quercetin using three different AMF strains and their combinations on grapes [48]. We were not able to confirm these results for okra in the presence of one AMF strain. But at the same time, the combined treatment of *F. mosseae* and *A. pullulans* strain DSM 14950 increased significantly the level of quercetin-3-diglucoside compared to other treatments. Furthermore, the individual application of *A. pullulans* strain DSM 14950 led to more elevated quercetin and quercetin-3-diglucoside contents.

As it is known, secondary metabolites in general are essential for the defense mechanism of plants and can be influenced by microorganisms, like AMF [49]. These compounds, released by both plants and AMF, act as signaling molecules during the symbiotic interaction between plants and AMF. Plants' responses to different stress factors cause secondary metabolism alterations, resulting in changes not only in polyphenol compounds but also in other metabolites, such as terpenoids [50,51]. Our GST enzyme activity measurements prove that the okra plants detected the presence of all the applied microorganisms as a slight biotic stress factor that resulted in elevated enzyme concentrations compared to the control treatment. The observed negative correlation between GST enzyme activity and coumaric acid indicates that the cause of this is the antioxidant activity of coumaric acid or the possibility that one monomer from the enzyme family of GST (4-CA) binds p-coumaric acid [52].

Our results highlight the potential benefits of utilizing these microorganisms in combination with mycorrhiza as a means of enhancing okra fruit phenolic content and productivity.

## 5. Conclusions

The aim of this study was to assess how the tested biocontrol microorganisms affect okra growth with a specific focus on the production of polyphenols in the fruit. Our preliminary results indicate that all tested microorganisms had various but generally positive effects on plant growth and yield, with the exception of *A. pullulans* strain DSM 14950 and *S. griseoviridis* strain K61 applied individually.

We were able to assign different microorganisms and their combinations to specific polyphenol compounds and determine the relationships between them. These results provide insights into how the content of specific polyphenol compounds can be selectively influenced through targeted inoculation with different microorganisms.

Our results highlight the potential of tested microorganisms in agricultural practices for improving crop quality and quantity. Further investigation in this field is warranted to fully elucidate the scope of their potential applications and optimize their utilization.

**Author Contributions:** Supervision, conceptualization, formal analysis, writing—original draft, writing—review and editing, resources, K.P. and G.T.; formal analysis, A.L.S.; formal analysis, A.I.A.Y.; writing, data curation, formal analysis, L.L.; data curation, formal analysis, investigation, methodology, methods, A.A.A.A. All authors have read and agreed to the published version of the manuscript.

**Funding:** The work was funded by Stipendium Hungaricum Scholarship ID: 253945. This research was funded by the National Research, Development and Innovation Office, grant number OTKA142974, and Agri-biotechnology and Precision Breeding for Food Security National Laboratory, grant number RRF-2.3.1-21-2022-00007.

**Institutional Review Board Statement:** This study did not involve human participants or animals. The current study complies with relevant institutional, national, and international guidelines and legislation for experimental research and field studies on plants (either cultivated or wild) and fungi, including the collection of plant and fungal materials.

**Data Availability Statement:** The associated dataset of this study is available upon request from the corresponding author.

**Conflicts of Interest:** The authors declare no conflicts of interest.

## Appendix A

**Table A1.** Results of one-way ANOVA.

| | F Value | P |
|---|---|---|
| Colonization | 25.223 | $2.643 \times 10^{-6}$ *** |
| Shoot FW | 27.161 | $9.537 \times 10^{-12}$ *** |
| Shoot DW | 29.249 | $3.523 \times 10^{-12}$ *** |
| Root FW | 21.455 | $2.083 \times 10^{-10}$ *** |
| Root DW | 10.937 | $5.755 \times 10^{-7}$ *** |
| Yield | 47.42 | $4.1 \times 10^{-15}$ *** |
| GST | 297.68 | $2.2 \times 10^{-16}$ *** |
| Kaempferol | 17.36 | $2.933 \times 10^{-9}$ *** |
| Coumaric acid | 33.488 | $5.545 \times 10^{-13}$ *** |
| Coumaroyle hexoside | 30.239 | $2.243 \times 10^{-12}$ *** |
| Quercetin | 202.9 | $<2.2 \times 10^{-16}$ *** |
| Quercetin-3-diglucoside | 422.88 | $<2.2 \times 10^{-16}$ *** |
| Chlorogenic acid | 17.245 | $3.181 \times 10^{-9}$ *** |
| Sinapoyl | 63.176 | $<2.2 \times 10^{-16}$ *** |
| Di-caffeyl quinic acid | 8.7901 | $5.327 \times 10^{-6}$ *** |

*** significant at $p < 0.001$.

**Table A2.** Results of Tukey's post hoc test (COL: colonization level; SFW: shoot fresh weight; SDW: shoot dry weight; RFW: root fresh weight; RDW: root dry weight; Y: yield; GST: gluthatione-S-transferase enzyme activity; KMP: kaempferol; CA: coumaric acid; CHex: coumaroyl hexoside; QCT: quercetin; Q3DG: quercetin-3-diglucoside; CGA: chlorogenic acid; SIN: sinapoyl DCQH: di-caffeyl quinic acid; C: no treatment; M: *F. mosseae*; T: *T. asperellum* strain T34; S: *S. griseoviridis* strain K61; A: *A. pullulans* strain DSM 14950; M + T: *F. mosseae* and *T. asperellum* strain T34; M + S: *F. mosseae* and *S. griseoviridis* strain K61; M + A: *F. mosseae* and *A. pullulans* strain DSM 14950).

| | COL | SFW | SDW | RFW | RDW | Y | GST | KMP | CA | CHex | QCT | QDG | CGA | SIN | DCQH |
|---|---|---|---|---|---|---|---|---|---|---|---|---|---|---|---|
| C − A | | 0.438 | 0.296 | 0.496 | 0.999 | 0.999 | 0.000 | 0.069 | 0.000 | 0.080 | 0.000 | 0.000 | 0.999 | 0.000 | 0.041 |
| M − A | | 0.000 | 0.000 | 0.000 | 0.001 | 0.000 | 0.999 | 0.072 | 0.000 | 0.815 | 0.000 | 0.000 | 0.000 | 0.999 | 0.037 |
| M + A − A | | 0.000 | 0.000 | 0.000 | 0.000 | 0.000 | 0.064 | 0.974 | 0.084 | 0.000 | 0.000 | 0.000 | 0.087 | 0.000 | 0.022 |
| M + S − A | | 0.005 | 0.009 | 0.000 | 0.011 | 0.000 | 0.000 | 0.353 | 0.000 | 0.137 | 0.000 | 0.000 | 0.064 | 0.000 | 0.031 |
| M + T − A | | 0.000 | 0.000 | 0.000 | 0.001 | 0.000 | 0.053 | 0.817 | 0.303 | 0.754 | 0.000 | 0.000 | 0.000 | 0.003 | 0.027 |
| S − A | | 0.999 | 0.995 | 0.999 | 0.999 | 0.000 | 0.062 | 0.381 | 0.092 | 0.999 | 0.000 | 0.000 | 0.056 | 0.999 | 0.003 |
| T − A | | 0.001 | 0.000 | 0.072 | 0.371 | 0.000 | 0.203 | 0.000 | 0.376 | 0.449 | 0.000 | 0.000 | 0.908 | 0.038 | 0.999 |
| M − C | | 0.000 | 0.000 | 0.001 | 0.002 | 0.000 | 0.000 | 0.000 | 0.983 | 0.088 | 0.000 | 0.673 | 0.000 | 0.000 | 0.008 |
| M + A − C | | 0.000 | 0.000 | 0.000 | 0.001 | 0.000 | 0.000 | 0.046 | 0.005 | 0.000 | 0.000 | 0.000 | 0.074 | 0.000 | 0.006 |
| M + S − C | | 0.467 | 0.752 | 0.016 | 0.019 | 0.000 | 0.000 | 0.000 | 0.000 | 0.000 | 0.000 | 0.027 | 0.066 | 0.648 | 0.022 |
| M + T − C | | 0.000 | 0.000 | 0.000 | 0.002 | 0.000 | 0.000 | 0.000 | 0.001 | 1.000 | 0.000 | 0.794 | 0.000 | 0.000 | 0.001 |
| S − C | | 0.789 | 0.735 | 0.299 | 0.999 | 0.000 | 0.000 | 0.000 | 0.023 | 0.568 | 0.000 | 0.048 | 0.055 | 0.000 | 0.000 |
| T − C | | 0.128 | 0.090 | 0.496 | 0.496 | 0.000 | 0.000 | 0.000 | 0.001 | 0.999 | 0.998 | 0.999 | 0.739 | 0.000 | 0.002 |
| M + A − M | 0.014 | 0.994 | 0.999 | 0.977 | 0.999 | 0.348 | 0.064 | 0.107 | 0.000 | 0.000 | 0.020 | 0.000 | 0.000 | 0.000 | 0.999 |
| M + S − M | 0.999 | 0.005 | 0.000 | 0.934 | 0.982 | 0.124 | 0.000 | 0.728 | 0.002 | 0.039 | 0.000 | 0.872 | 0.007 | 0.000 | 0.003 |
| M + T − M | 0.000 | 0.999 | 0.998 | 0.999 | 1.000 | 0.051 | 0.071 | 0.272 | 0.000 | 0.075 | 0.000 | 0.999 | 0.025 | 0.001 | 0.999 |
| S − M | | 0.000 | 0.000 | 0.000 | 0.000 | 0.999 | 0.056 | 0.999 | 0.002 | 0.953 | 0.000 | 0.921 | 0.002 | 1.000 | 0.000 |
| T − M | | 0.036 | 0.006 | 0.112 | 0.214 | 0.999 | 0.086 | 0.025 | 0.000 | 0.071 | 0.002 | 0.840 | 0.000 | 0.494 | 0.031 |
| M + S − M + A | 0.010 | 0.001 | 0.000 | 0.419 | 0.935 | 0.000 | 0.224 | 0.905 | 0.000 | 0.000 | 0.000 | 0.000 | 0.986 | 0.000 | 0.000 |
| M + T − M + A | 0.057 | 0.999 | 0.999 | 0.999 | 0.999 | 0.000 | 0.636 | 0.999 | 0.997 | 0.000 | 0.001 | 0.000 | 0.025 | 0.000 | 0.988 |
| S − M + A | | 0.000 | 0.000 | 0.000 | 0.000 | 0.140 | 0.481 | 0.287 | 0.999 | 0.000 | 0.000 | 0.000 | 0.999 | 0.000 | 0.000 |
| T − M + A | | 0.005 | 0.011 | 0.056 | 0.132 | 0.662 | 0.601 | 0.000 | 0.991 | 0.000 | 0.000 | 0.000 | 0.448 | 0.000 | 0.016 |
| M + T − M + S | 0.000 | 0.001 | 0.001 | 0.782 | 0.977 | 0.336 | 0.000 | 0.993 | 0.000 | 0.000 | 0.000 | 0.770 | 0.023 | 0.000 | 0.001 |
| S − M + S | | 0.023 | 0.042 | 0.000 | 0.004 | 0.218 | 0.000 | 0.949 | 0.000 | 0.034 | 0.803 | 0.999 | 0.999 | 0.000 | 0.916 |
| T − M + S | | 0.994 | 0.857 | 0.686 | 0.731 | 0.054 | 0.000 | 0.006 | 0.000 | 0.000 | 0.000 | 0.137 | 0.094 | 0.000 | 0.008 |
| S − M + T | | 0.000 | 0.000 | 0.000 | 0.000 | 0.053 | 0.978 | 0.569 | 0.919 | 0.520 | 0.001 | 0.840 | 0.031 | 0.001 | 0.000 |
| T − M + T | | 0.010 | 0.033 | 0.050 | 0.198 | 0.051 | 0.727 | 0.001 | 0.999 | 0.999 | 0.000 | 0.921 | 0.002 | 0.110 | 0.005 |
| T − S | | 0.003 | 0.001 | 0.054 | 0.198 | 0.992 | 0.093 | 0.033 | 0.866 | 0.250 | 0.000 | 0.179 | 0.202 | 0.046 | 0.001 |

**Table A3.** Pearson correlation coefficient values (R: Pearson correlation; P: *p*-value; KMP: kaempferol; CA: coumaric acid; CHex: coumaroyl hexoside; QCT: quercetin; Q3DG: quercetin-3-diglucoside; CGA: chlorogenic acid; SIN: sinapoyl DCQH: di-caffeyl quinic acid; SFW: shoot fresh weight; SDW: shoot dry weight; RFW: root fresh weight; RDW: root dry weight; Y: yield; GST: gluthation-S-transferase enzyme activity). * moderate degree of correlation at R > ±0.30, ** high degree of correlation at R > ±0.50.

| | | KMP | CA | CHex | QCT | QDG | CGA | SIN | DCQH | SFW | SDW | RFW | RDW | Y | GST |
|---|---|---|---|---|---|---|---|---|---|---|---|---|---|---|---|
| KMP | R | 1.00 | −0.07 | 0.01 | −0.23 | −0.26 | 0.42 ** | 0.28 | −0.02 | 0.17 | 0.24 | 0.19 | 0.23 | 0.51 ** | 0.57 ** |
| | P | | 0.66 | 0.97 | 0.15 | 0.10 | 0.01 | 0.08 | 0.90 | 0.29 | 0.13 | 0.24 | 0.14 | 0.00 | 0.00 |
| CA | R | −0.07 | 1.00 | 0.13 | −0.45 ** | −0.52 ** | 0.28 | −0.58 ** | −0.22 | 0.06 | 0.07 | 0.22 | 0.21 | 0.24 | −0.53 ** |
| | P | 0.66 | | 0.42 | 0.00 | 0.00 | 0.08 | 0.00 | 0.18 | 0.70 | 0.69 | 0.17 | 0.18 | 0.14 | 0.00 |
| CHex | R | 0.01 | 0.13 | 1.00 | −0.27 | 0.18 | 0.19 | 0.41 ** | 0.23 | 0.39 * | 0.36 * | 0.42 ** | 0.40 ** | 0.07 | −0.13 |
| | P | 0.97 | 0.42 | | 0.09 | 0.26 | 0.25 | 0.01 | 0.16 | 0.01 | 0.02 | 0.01 | 0.01 | 0.69 | 0.43 |
| QCT | R | −0.23 | −0.45 ** | −0.27 | 1.00 | 0.66 ** | −0.52 ** | −0.02 | 0.31 | −0.23 | −0.21 | −0.27 | −0.31 | −0.75 ** | 0.24 |
| | P | 0.15 | 0.00 | 0.09 | | 0.00 | 0.00 | 0.90 | 0.05 | 0.14 | 0.19 | 0.09 | 0.05 | 0.00 | 0.13 |
| Q3DG | R | −0.26 | −0.52 ** | 0.18 | 0.66 ** | 1.00 | −0.41 ** | 0.27 | 0.21 | −0.29 | −0.30 | −0.28 | −0.22 | −0.61 ** | 0.29 |
| | P | 0.10 | 0.00 | 0.26 | 0.00 | | 0.01 | 0.09 | 0.20 | 0.07 | 0.06 | 0.08 | 0.17 | 0.00 | 0.07 |
| CGA | R | 0.42 ** | 0.28 | 0.19 | −0.52 ** | −0.41 ** | 1.00 | 0.24 | 0.14 | 0.50 ** | 0.58 ** | 0.51 ** | 0.56 ** | 0.61 ** | 0.27 |
| | P | 0.01 | 0.08 | 0.25 | 0.00 | 0.01 | | 0.14 | 0.39 | 0.00 | 0.00 | 0.00 | 0.00 | 0.00 | 0.10 |
| SIN | R | 0.28 | −0.58 ** | 0.41 ** | −0.02 | 0.27 | 0.24 | 1.00 | 0.47 ** | 0.56 ** | 0.52 ** | 0.38 * | 0.38 * | 0.25 | 0.48 ** |
| | P | 0.08 | 0.00 | 0.01 | 0.90 | 0.09 | 0.14 | | 0.00 | 0.00 | 0.00 | 0.01 | 0.01 | 0.13 | 0.00 |
| DCQH | R | −0.02 | −0.22 | 0.23 | 0.31 | 0.21 | 0.14 | 0.47 ** | 1.00 | 0.58 ** | 0.61 ** | 0.45 ** | 0.41 ** | 0.02 | 0.21 |
| | P | 0.90 | 0.18 | 0.16 | 0.05 | 0.20 | 0.39 | 0.00 | | 0.00 | 0.00 | 0.00 | 0.01 | 0.92 | 0.19 |
| SFW | R | 0.17 | 0.06 | 0.39 * | −0.23 | −0.29 | 0.50 ** | 0.56 ** | 0.58 ** | 1.00 | 0.89 ** | 0.83 ** | 0.76 ** | 0.51 ** | 0.13 |
| | P | 0.29 | 0.70 | 0.01 | 0.14 | 0.07 | 0.00 | 0.00 | 0.00 | | 0.00 | 0.00 | 0.00 | 0.00 | 0.42 |
| SDW | R | 0.24 | 0.07 | 0.36 * | −0.21 | −0.30 | 0.58 ** | 0.52 ** | 0.61 ** | 0.89 ** | 1.00 | 0.80 ** | 0.68 ** | 0.47 ** | 0.16 |
| | P | 0.13 | 0.69 | 0.02 | 0.19 | 0.06 | 0.00 | 0.00 | 0.00 | 0.00 | | 0.00 | 0.00 | 0.00 | 0.32 |
| RFW | R | 0.19 | 0.22 | 0.42 ** | −0.27 | −0.28 | 0.51 ** | 0.38 * | 0.45 ** | 0.83 ** | 0.80 ** | 1.00 | 0.75 ** | 0.54 ** | −0.02 |
| | P | 0.24 | 0.17 | 0.01 | 0.09 | 0.08 | 0.00 | 0.01 | 0.00 | 0.00 | 0.00 | | 0.00 | 0.00 | 0.90 |
| RDW | R | 0.23 | 0.21 | 0.40 ** | −0.31 | −0.22 | 0.56 ** | 0.38 * | 0.41 ** | 0.76 ** | 0.68 ** | 0.75 ** | 1.00 | 0.46 ** | 0.10 |
| | P | 0.14 | 0.18 | 0.01 | 0.05 | 0.17 | 0.00 | 0.01 | 0.01 | 0.00 | 0.00 | 0.00 | | 0.00 | 0.55 |
| Y | R | 0.51 ** | 0.24 | 0.07 | −0.75 ** | −0.61 ** | 0.61 ** | 0.25 | 0.02 | 0.51 ** | 0.47 ** | 0.54 ** | 0.46 ** | 1.00 | 0.14 |
| | P | 0.00 | 0.14 | 0.69 | 0.00 | 0.00 | 0.00 | 0.13 | 0.92 | 0.00 | 0.00 | 0.00 | 0.00 | | 0.39 |
| GST | R | 0.57 ** | −0.53 ** | −0.13 | 0.24 | 0.29 | 0.27 | 0.48 ** | 0.21 | 0.13 | 0.16 | −0.02 | 0.10 | 0.14 | 1.00 |
| | P | 0.00 | 0.00 | 0.43 | 0.13 | 0.07 | 0.10 | 0.00 | 0.19 | 0.42 | 0.32 | 0.90 | 0.55 | 0.39 | |

**Table A4.** Eigenvalues of the principal component analysis.

| | PC1 | PC2 | PC3 | PC4 | PC5 | PC6 | PC7 | PC8 |
|---|---|---|---|---|---|---|---|---|
| Eigenvalue | 2.682 | 2.094 | 1.170 | 0.832 | 0.551 | 0.408 | 0.136 | 0.127 |
| Variability (%) | 33.529 | 26.181 | 14.620 | 10.406 | 6.882 | 5.096 | 1.697 | 1.590 |
| Cumulative % | 33.529 | 59.710 | 74.330 | 84.736 | 91.617 | 96.713 | 98.410 | 100.000 |

**Table A5.** Correlations between variables and factors of the principal component analysis.

| | PC1 | PC2 | PC3 | PC4 | PC5 | PC6 | PC7 | PC8 |
|---|---|---|---|---|---|---|---|---|
| Kaempferol | 0.186 | 0.350 | −0.571 | −0.117 | −0.617 | 0.130 | 0.098 | 0.110 |
| Coumaric acid | 0.455 | −0.183 | 0.407 | 0.254 | −0.371 | 0.051 | 0.450 | −0.433 |
| Coumaroyl hexoside | 0.012 | 0.394 | 0.651 | −0.341 | −0.268 | 0.198 | −0.436 | 0.058 |
| Quercetin | −0.504 | −0.189 | −0.111 | 0.312 | −0.405 | −0.063 | −0.440 | −0.489 |
| Quercetin-3-diglucoside | −0.519 | 0.014 | 0.173 | −0.233 | −0.367 | −0.446 | 0.498 | 0.255 |
| Chlorogenic acid | 0.352 | 0.414 | −0.066 | 0.269 | −0.062 | −0.756 | −0.228 | 0.017 |
| Sinapoyl | −0.221 | 0.587 | −0.072 | −0.171 | 0.322 | 0.037 | 0.286 | −0.621 |
| Di-caffeylquinic acid | −0.249 | 0.368 | 0.179 | 0.743 | 0.058 | 0.291 | 0.151 | 0.328 |

**Table A6.** Peak identification of the HPLC profile of phenolic compounds.

| Peak No. | Rt | Absorption Maxima (nm) | | | | Compound's Name |
|---|---|---|---|---|---|---|
| 1 | 7.372 | 297 | 326 | | | Feruloyl dihexose |
| 2 | 8.692 | 281 | 313 | | | Coumaroyl dihexose |
| 3 | 9.211 | 281 | 300 | 328 | | Sinapoyl di-glucoside |
| 4 | 9.803 | 299 | 325 | | | Feruloyl hexose |
| 5 | 10.202 | 216 | 242 | 299 | 331 | Sinapoyl hexose |
| 6 | 10.854 | 279 | 302 | 331 | | Sinapoyl catechoyl derivative |
| 7 | 11.714 | 302 | 329 | | | Sinapoyl derivative |
| 8 | 12.801 | 289 | 312 | | | Coumaroyl glucoside |
| 9 | 13.116 | 210 | 230 | 314 | | Caffeoylquinic acid derivative |
| 10 | 14.442 | 311 | | | | Coumaric acid |
| 11 | 14.843 | 216 | 238 | 297 (sh) | 330 | Sinapic acid derivative |
| 12 | 15.982 | 298 | 327 | | | Neochlorogenic acid |
| 13 | 16.311 | 284 | 328 | | | Chlorogenic acid isomer |
| 14 | 16.643 | 294 | 327 | | | Chlorogenic acid |
| 15 | 17.341 | 300 | 330 | | | Sinapic acid |
| 16 | 17.764 | 282 | 326 | | | Ferulic acid derivative |
| 17 | 18.132 | 282 | 325 | | | Feruloyl catechin derivative |
| 18 | 18.442 | 282 | 322 | | | Caffeic acid |
| 19 | 20.053 | 282 | 311 | | | Di-coumaroylquinic acid-glucose |
| 20 | 20.69 | 255 | 266 | 354 | | Quercetin.3-o-glucose-xylose |
| 21 | 21.471 | 255 | 266 | 318 | 362 | Isorahmnetin-3-coffeoly di-glucoside |
| 22 | | 255 | | | | |
| 23 | 24.317 | 297 | 329 | | | Di-coffeoylquinic acid |
| 24 | 25.634 | | | | | |
| 25 | 27.482 | 255 | 266 | 355 | | Rutin |
| 26 | 28.227 | 268 | 348 | | | Kaempferoyl-3-glucoside |
| 27 | 29.416 | 222 | 266 | 355 | | Quercetin-3-glucoside |
| 28 | 30.112 | | | | | |
| 29 | 30.604 | 298 | 332 | | | Miric |
| 30 | 36.354 | 241 | 297 | 322 | | Di-coffeoyl derivative |

## References


1. Jain, N.; Jain, R.; Jain, V.; Jain, S. A review on Abelmoschus esculentus. *Pharmacia* **2012**, *1*, 84–89. Available online: https://www.pharmacia.ipsgwalior.org/artical/vol1_issue3_2.pdf (accessed on 11 April 2024).
2. Wahyuningsih, S.P.A.; Savira, N.I.I.; Anggraini, D.W.; Winarni, D.; Suhargo, L.; Kusuma, B.K.A.; Nindyasari, F.; Setianingsih, N.; Mwendolwa, A.A. Antioxidant and Nephroprotective Effects of Okra Pods Extract (*Abelmoschus esculentus* L.) against Lead Acetate-Induced Toxicity in Mice. *Scientifica* **2020**, *2020*, 4237205. [CrossRef]
3. Kumar, A.; Kumar, P.; Nadendla, R. A review on: *Abelmoschus esculentus* (Okra). *Int. Res. J. Pharm. Appl. Sci.* **2013**, *3*, 129–132. Available online: https://d1wqtxts1xzle7.cloudfront.net/38521605/IRJPAS_34129-132_1-libre.pdf?1440050475=&response-content-disposition=inline;+filename=A_REVIEW_ON_ABELMOSCHUS_ESCULENTUS_OKRA.pdf&Expires=1715083412&Signature=SERYkKd-N6QivBNsLzS~~cw5Owi92i-OPL87RW4okXjf (accessed on 11 April 2024).
4. Yan, W.; Shi, Y.-X.; Chai, A.; Xie, X.; Guo, M.; Li, B. Verticillium Wilt of Okra Caused by *Verticillium dahliae* Kleb. in China. *Mycobiology* **2018**, *46*, 254–259. [CrossRef]
5. Elkhalifa, A.E.O.; Alshammari, E.; Adnan, M.; Alcantara, J.C.; Awadelkareem, A.M.; Eltoum, N.E.; Mehmood, K.; Panda, B.P.; Ashraf, S.A. Okra (*Abelmoschus esculentus*) as a Potential Dietary Medicine with Nutraceutical Importance for Sustainable Health Applications. *Molecules* **2021**, *26*, 696. [CrossRef] [PubMed]
6. Singh, J.; Nigam, R. Importance of Okra (*Abelmoschus esculentus* L.) and It's Proportion in the World as a Nutritional Vegetable. *Int. J. Environ. Clim. Chang.* **2023**, *13*, 1694–1699. [CrossRef]
7. Lockyer, S.; White, A.; Buttriss, J.L. Biofortified crops for tackling micronutrient deficiencies—What impact are these having in developing countries and could they be of relevance within Europe? *Nutr. Bull.* **2018**, *43*, 319–357. [CrossRef]
8. Tulchinsky, T.H. Micronutrient Deficiency Conditions: Global Health Issues. *Public Health Rev.* **2010**, *32*, 243–255. [CrossRef]
9. Moyin-Jesu, E.I. Use of plant residues for improving soil fertility, pod nutrients, root growth and pod weight of okra (*Abelmoschus esculentum* L.). *Bioresour. Technol.* **2007**, *98*, 2057–2064. [CrossRef]

10. Sami, R.; Li, Y.; Qi, B.; Wang, S.; Zhang, Q.; Han, F.; Ma, Y.; Jing, J.; Jiang, L. HPLC Analysis of Water-Soluble Vitamins (B2, B3, B6, B12, and C) and Fat-Soluble Vitamins (E, K, D, A, and β-Carotene) of Okra (*Abelmoschus esculentus*). *J. Chem.* **2014**, *2014*, 831357. [CrossRef]
11. Silaghi-Dumitrescu, R.; Oana (Gadina), F.-M.; Lehene, M. Antioxidants in Clinical Treatments. In *Fundamental and Biomedical Aspects of Redox*; IGI Global: Hershey, PA, USA, 2023; pp. 315–326. [CrossRef]
12. Niki, E. Antioxidant Defense Network and Vitamin E. In *Food Chemistry, Function and Analysis*; The Royal Society of Chemistry: Croydon, UK, 2019; pp. 134–150. [CrossRef]
13. Wu, D.-T.; Nie, X.-R.; Shen, D.-D.; Li, H.-Y.; Zhao, L.; Zhang, Q.; Lin, D.-R.; Qin, W. Phenolic Compounds, Antioxidant Activities, and Inhibitory Effects on Digestive Enzymes of Different Cultivars of Okra (*Abelmoschus esculentus*). *Molecules* **2020**, *25*, 1276. [CrossRef]
14. Ji, M.; Gong, X.; Li, X.; Wang, C.; Li, M. Advanced Research on the Antioxidant Activity and Mechanism of Polyphenols from Hippophae Species—A Review. *Molecules* **2020**, *25*, 917. [CrossRef] [PubMed]
15. Handique, J.G.; Baruah, J.B. Polyphenolic compounds: An overview. *React. Funct. Polym.* **2002**, *52*, 163–188. [CrossRef]
16. Yahfoufi, N.; Alsadi, N.; Jambi, M.; Matar, C. The Immunomodulatory and Anti-Inflammatory Role of Polyphenols. *Nutrients* **2018**, *10*, 1618. [CrossRef] [PubMed]
17. Niedzwiecki, A.; Roomi, M.; Kalinovsky, T.; Rath, M. Anticancer Efficacy of Polyphenols and Their Combinations. *Nutrients* **2016**, *8*, 552. [CrossRef] [PubMed]
18. Quiñones, M.; Miguel, M.; Aleixandre, A. Beneficial effects of polyphenols on cardiovascular disease. *Pharmacol. Res.* **2013**, *68*, 125–131. [CrossRef]
19. Daglia, M. Polyphenols as antimicrobial agents. *Curr. Opin. Biotechnol.* **2012**, *23*, 174–181. [CrossRef]
20. Stürmer, S.L.; Bever, J.D.; Morton, J.B. Biogeography of arbuscular mycorrhizal fungi (Glomeromycota): A phylogenetic perspective on species distribution patterns. *Mycorrhiza* **2018**, *28*, 587–603. [CrossRef]
21. Noceto, P.-A.; Bettenfeld, P.; Boussageon, R.; Hériché, M.; Sportes, A.; van Tuinen, D.; Courty, P.-E.; Wipf, D. Arbuscular mycorrhizal fungi, a key symbiosis in the development of quality traits in crop production, alone or combined with plant growth-promoting bacteria. *Mycorrhiza* **2021**, *31*, 655–669. [CrossRef]
22. Guzmán-Guzmán, P.; Kumar, A.; Santos-Villalobos, S.d.L.; Parra-Cota, F.I.; Orozco-Mosqueda, M.d.C.; Fadiji, A.E.; Hyder, S.; Babalola, O.O.; Santoyo, G. Trichoderma Species: Our Best Fungal Allies in the Biocontrol of Plant Diseases—A Review. *Plants* **2023**, *12*, 432. [CrossRef]
23. Doumbou, C.L.; Salove, M.K.H.; Crawford, D.L.; Beaulieu, C. Actinomycetes, promising tools to control plant diseases and to promote plant growth. *Phytoprotection* **2005**, *82*, 85–102. [CrossRef]
24. Di Francesco, A.; Zajc, J.; Stenberg, J.A. *Aureobasidium* spp.: Diversity, Versatility, and Agricultural Utility. *Horticulturae* **2023**, *9*, 59. [CrossRef]
25. Feldmann, F.; Idczak, E. 18 Inoculum Production of Vesicular-arbuscular Mycorrhizal Fungi for Use in Tropical Nurseries. *Methods Microbiol.* **1992**, *24*, 339–357. [CrossRef]
26. Vierheilig, H.; Coughlan, A.P.; Wyss, U.; Piche, Y. Ink and Vinegar, a Simple Staining Technique for Arbuscular-Mycorrhizal Fungi. *Appl. Environ. Microbiol.* **1998**, *64*, 5004–5007. [CrossRef] [PubMed]
27. Giovannetti, M.; Mosse, B. An Evaluation of Techniques for Measuring Vesicular Arbuscular Mycorrhizal Infection in Roots. *New. Phytol.* **1980**, *84*, 489–500. [CrossRef]
28. Habig, W.H.; Pabst, M.J.; Jakoby, W.B. Glutathione S-transferases. The first enzymatic step in mercapturic acid formation. *J. Biol. Chem.* **1974**, *249*, 7130–7139. [CrossRef] [PubMed]
29. Bradford, M.M. A rapid and sensitive method for the quantitation of microgram quantities of protein utilizing the principle of protein-dye binding. *Anal. Biochem.* **1976**, *72*, 248–254. [CrossRef] [PubMed]
30. Romdhane, M.H.; Chahdoura, H.; Barros, L.; Dias, M.I.; Corrêa, R.C.G.; Morales, P.; Ciudad-Mulero, M.; Flamini, G.; Majdoub, H.; Ferreira, I.C.F.R. Chemical Composition, Nutritional Value, and Biological Evaluation of Tunisian Okra Pods (*Abelmoschus esculentus* L. Moench). *Molecules* **2020**, *25*, 4739. [CrossRef] [PubMed]
31. Yang, J.; Chen, X.; Rao, S.; Li, Y.; Zang, Y.; Zhu, B. Identification and Quantification of Flavonoids in Okra (*Abelmoschus esculentus* L. Moench) and Antiproliferative Activity In Vitro of Four Main Components Identified. *Metabolites* **2022**, *12*, 483. [CrossRef]
32. Wang, X.; Liu, X.; Shi, N.; Zhang, Z.; Chen, Y.; Yan, M.; Li, Y. Response surface methodology optimization and HPLC-ESI-QTOF-MS/MS analysis on ultrasonic-assisted extraction of phenolic compounds from okra (*Abelmoschus esculentus*) and their antioxidant activity. *Food Chem.* **2023**, *405*, 134966. [CrossRef]
33. Zeng, Y.; Guo, L.-P.; Chen, B.-D.; Hao, Z.-P.; Wang, J.-Y.; Huang, L.-Q.; Yang, G.; Cui, X.-M.; Yang, L.; Wu, Z.-X.; et al. Arbuscular mycorrhizal symbiosis and active ingredients of medicinal plants: Current research status and prospective. *Mycorrhiza* **2013**, *23*, 253–265. [CrossRef] [PubMed]
34. Ali, M.M.; Sani, M.N.; Arifunnahar, M.A.M.; Aminuzzaman, F.M.; Mridha, M.A.U. Influence of arbuscular mycorrhizal fungi on growth, nutrient uptake and disease suppression of some selected vegetable crops. *Azarian J. Agric.* **2018**, *5*, 190–196.
35. Mwangi, M.W.; Monda, E.O.; Okoth, S.A.; Jefwa, J.M. Inoculation of tomato seedlings with Trichoderma Harzianum and Arbuscular Mycorrhizal Fungi and their effect on growth and control of wilt in tomato seedlings. *Braz. J. Microbiol.* **2011**, *42*, 508–513. [CrossRef] [PubMed]

36. Ndoye, F.; Kane, A.; Bakhoum, N.; Sanon, A.; Fall, D.; Diouf, D.; Sylla, S.N.; Bâ, A.M.; Sy, M.O.; Noba, K. Response of *Acacia senegal* (L.) Willd. to inoculation with arbuscular mycorrhizal fungi isolates in sterilized and unsterilized soils in Senegal. *Agrofor. Syst.* **2013**, *87*, 941–952. [CrossRef]
37. Frey-Klett, P.; Garbaye, J.; Tarkka, M. The mycorrhiza helper bacteria revisited. *New Phytol.* **2007**, *176*, 22–36. [CrossRef] [PubMed]
38. Hoeksema, J.D.; Chaudhary, V.B.; Gehring, C.A.; Johnson, N.C.; Karst, J.; Koide, R.T.; Pringle, A.; Zabinski, C.; Bever, J.D.; Moore, J.C.; et al. A meta-analysis of context-dependency in plant response to inoculation with mycorrhizal fungi. *Ecol. Lett.* **2010**, *13*, 394–407. [CrossRef] [PubMed]
39. Stewart, A.; Hill, R. Applications of Trichoderma in Plant Growth Promotion. In *Biotechnology and Biology of Trichoderma*; Elsevier: Amsterdam, The Netherlands, 2014; pp. 415–428. [CrossRef]
40. Patel, J.; Teli, B.; Bajpai, R.; Meher, J.; Rashid, M.; Mukherjee, A.; Yadav, S.K. Trichoderma-mediated biocontrol and growth promotion in plants: An endophytic approach. In *Role of Plant Growth Promoting Microorganisms in Sustainable Agriculture and Nanotechnology*; Elsevier: Amsterdam, The Netherlands, 2019; pp. 219–239. [CrossRef]
41. Cutrim, C.S.; Cortez, M.A.S. A review on polyphenols: Classification, beneficial effects and their application in dairy products. *Int. J. Dairy Technol.* **2018**, *71*, 564–578. [CrossRef]
42. Fabianová, J.; Šlosár, M.; Kopta, T.; Vargová, A.; Timoracká, M.; Mezeyová, I.; Andrejiová, A. Yield, Antioxidant Activity and Total Polyphenol Content of Okra Fruits Grown in Slovak Republic. *Horticulturae* **2022**, *8*, 966. [CrossRef]
43. Xia, F.; Zhong, Y.; Li, M.; Chang, Q.; Liao, Y.; Liu, X.; Pan, R. Antioxidant and Anti-Fatigue Constituents of Okra. *Nutrients* **2015**, *7*, 8846–8858. [CrossRef]
44. Chen, S.; Jin, W.; Liu, A.; Zhang, S.; Liu, D.; Wang, F.; Lin, X.; He, C. Arbuscular mycorrhizal fungi (AMF) increase growth and secondary metabolism in cucumber subjected to low temperature stress. *Sci. Hortic.* **2013**, *160*, 222–229. [CrossRef]
45. Castellanos-Morales, V.; Villegas, J.; Wendelin, S.; Vierheilig, H.; Eder, R.; Cárdenas-Navarro, R. Root colonisation by the arbuscular mycorrhizal fungus Glomus intraradices alters the quality of strawberry fruits (*Fragaria* × *ananassa* Duch.) at different nitrogen levels. *J. Sci. Food Agric.* **2010**, *90*, 1774–1782. [CrossRef]
46. Weng, W.; Yan, J.; Zhou, M.; Yao, X.; Gao, A.; Ma, C.; Cheng, J.; Ruan, J. Roles of Arbuscular mycorrhizal Fungi as a Biocontrol Agent in the Control of Plant Diseases. *Microorganisms* **2022**, *10*, 1266. [CrossRef] [PubMed]
47. Eftekhari, M.; Alizadeh, M.; Ebrahimi, P. Evaluation of the total phenolics and quercetin content of foliage in mycorrhizal grape (*Vitis vinifera* L.) varieties and effect of postharvest drying on quercetin yield. *Ind. Crops Prod.* **2012**, *38*, 160–165. [CrossRef]
48. Kaur, S.; Suseela, V. Unraveling arbuscular mycorrhiza-induced changes in plant primary and secondary metabolome. *Metabolites* **2020**, *10*, 335. [CrossRef] [PubMed]
49. Al-Khayri, J.M.; Rashmi, R.; Toppo, V.; Chole, P.B.; Banadka, A.; Sudheer, W.N.; Nagella, P.; Shehata, W.F.; Al-Mssallem, M.Q.; Alessa, F.M.; et al. Plant Secondary Metabolites: The Weapons for Biotic Stress Management. *Metabolites* **2023**, *13*, 716. [CrossRef] [PubMed]
50. Perincherry, L.; Lalak-Kańczugowska, J.; Stępień, Ł. Fusarium-Produced Mycotoxins in Plant-Pathogen Interactions. *Toxins* **2019**, *11*, 664. [CrossRef] [PubMed]
51. Pei, K.; Ou, J.; Huang, J.; Ou, S. Coumaric acid and its conjugates: Dietary sources, pharmacokinetic properties and biological activities. *J. Sci. Food Agric.* **2016**, *96*, 2952–2962. [CrossRef]
52. Dean, J.V.; Devarenne, T.P.; Lee, I.S.; Orlofsky, L.E. Properties of a Maize Glutathione S-Transferase That Conjugates Coumaric Acid and Other Phenylpropanoids. *Plant Physiol.* **1995**, *108*, 985–994. [CrossRef]

MDPI AG
Grosspeteranlage 5
4052 Basel
Switzerland
Tel.: +41 61 683 77 34

*Agriculture* Editorial Office
E-mail: agriculture@mdpi.com
www.mdpi.com/journal/agriculture

www.ingramcontent.com/pod-product-compliance
Lightning Source LLC
Chambersburg PA
CBHW062050220926
51850CB00010B/2723
* 9 7 8 3 7 2 5 8 6 3 6 0 0 *